Diagnosis: Rare Disease

By Denise Crompton

AllStar
P R E S S

Published by All Star Press - *Books that Change Lives*

Edited by Shaunee Michaud, Richard J. Nilsen

First edition: October 2014

Second edition: August 2015

ISBN #978-1-937376-17-8

Library of Congress number 2014950524

WHAT OTHERS ARE SAYING ABOUT
"DIAGNOSIS: RARE DISEASE"

"This book gives a heartbreaking exposé of what life is like for families that have a member with a rare disease. It follows the journey of 13 families who have one or more children with a rare disease called Mucolipidosis, as they wade through the erratic diagnosis process, medical muck-ups, endless pain, a plethora of medical appliances and home modifications, special education, multiple hospitalizations and surgeries, unrelenting stress and the ever-present fear of death.

The book highlights how stoic, resilient, strong and resourceful these families are, and how heartless, or thoughtless, some medical professionals can be.

It underlines the importance of networks to allow these isolated families to support each other and to give them a sense of belonging.

Denise Crompton presents these stories in a style that is compassionate and easy to read. Other families with a rare disease will recognize themselves in this book. It is a must for medical professionals."

~ Carolyn Paisley-Dew, ISMRD Board Member, Australia

"Regarding the topic of Diagnosis: Rare Disease, Denise Crompton amply demonstrates two of her important assets: one is the personal experience with ML III in her own daughter and two, her manifest ability to widen the subject and to bring a streamlined report on a number of other patients and families with either ML II or ML III. **The book is to become a classic in the still little-known world of very rare or orphan hereditary disorders.**"

~ JG Leroy MD. PhD

Professor & Chairman Emeritus

Depts.Pediatrics & Med. Genetics

Univ. Hosp. & Univ. Ghent

GHENT, Belgium

Senior Research Scholar

Greenwood Genetic Center

Greenwood, South Carolina

"Wow... breadth & depth. The various chapters of *Diagnosis: Rare Disease* cover every aspect of the impact of a rare disease on individuals. Within each chapter is deep insight into the tremendous differences of impact on each of 15 Individuals. In addition the style is compelling reading."

~ Bob Gorman - Independent Research Professional

"Denise Crompton captures the essence of living with a rare disease; the quest for a diagnosis, the fears, frustrations, heartbreaks, navigating the medical establishment, and the loss of loved ones. As one who has children battling a rare disease, I am encouraged by the journeys of the individual families Denise writes about, knowing that I am not alone in my feelings and struggles, and energized by the families' determination to live their lives to the fullest."

~ Susan Anganes

"As a retired operating room nurse, I have been amazed at the history of Kelley Crompton. During her lifetime she underwent numerous surgeries in hopes of improving her life style, preserving the same and hoping for a cure. I know how tough it is to enter an operating room the first time but returning again and again can be demoralizing. I can tell you firsthand how those giving such care are frustrated beyond belief. We are there to help and hope for lasting remedies. To see the same patient return numerous times is heart wrenching. My hope is that all health care givers will read this book so that the care they give will be compassionate. Sometimes just being there, truly listening and supporting a patient is all we can do. But we must start!"

~ Jeanne C. Sturrock, retired operating room nurse

DEDICATION

In loving memory of my daughter Kelley Crompton

And

In dedication to all who are affected with rare diseases.

ACKNOWLEDGEMENTS

I would like to express my sincere gratitude to the many family members and friends who have offered support, understanding and prayers for our daughter Kelley through the years, as well as the priests who have offered Masses and counsel.

It would be impossible for me to completely express the appreciation I have of the many members of the medical field we have encountered. Most of them have been humble, dedicated, and caring individuals. Although I would like to pay a personal tribute to all of them, those who have been the most instrumental deserve a mention by name, in the order in which we met them: Doctors John Wallace Zeller, Herbert L. Coffin, Murray Feingold, Thaddeus Kelly, Michael Goldberg, Kenneth Weinbeck, William Donaldson, Heinrich Wurm, Hermes Grillo, Cameron D. Wright, Mark J. Aronson, Eliza A. Deery, Gilbert J. Fanciullo, William C. Nugent, Peter A. DeLong, Sara S. Cathey, Jules G. Leroy, and Richard Wood, RN.

To say that my family is thankful for the help they have provided is an understatement. These fine people kept Kelley alive, and as comfortable as possible, for 45 years. They gave her parents hope and encouragement.

Father Gary Kosmowski's spiritual counsel to Kelley during her final years brought comfort to her and her grateful parents.

It is important for me to mention the National Organization of Rare Diseases (NORD): www.rarediseases.org.

It was through NORD that I learned of the National MPS Society: www.mpssociety.org.

And through the National MPS Society we were able to join up with the ISMRD: www.ismrd.org

These organizations have supplied us with connections to others who are dealing with rare diseases, and the information we have gained has been invaluable.

This book would not have been possible without the graciousness of these wonderful mothers who have openly shared the experiences they have had as parents of children with rare diseases, Pam Tobey, Trish Dennis, Jenny Noble, Liz Anthony, Debbie Nagle, Terri Klein, Jane Andrews, Linda Barham, Jackie James, Andrea Gates, Maria Elena Cardenas and Brenda Haggett.

I must also add a special thanks to Shaunee Michaud for her editing assistance. Finally, thank you, Rich Nilsen! I appreciate the fact that you saw the value in this project. It has been a pleasure to work with you to have this book published by *All Star Press*.

Denise Crompton

CONTENTS

FOREWARD I

Before encountering my first patient diagnosed with Mucolipidosis (ML) I had graduated from medical school, completed my first residency in pediatrics, practiced general pediatrics for several years, and was well into my second residency in medical genetics. My mentor, Dr. Jules Leroy, then a Visiting Senior Scholar at Greenwood Genetic Center where I was the clinical genetics trainee, had first described ML in the medical literature forty years earlier. It was 2005, the causative gene had only recently been described, and a couple brought their affected young daughter to Greenwood to meet Dr. Leroy. This visit prompted a laboratory project to identify disease causing mutations in the gene. The child's parents were involved with the patient advocacy group known as ISMRD. When other ISRMD families learned there were doctors in Greenwood interested in ML, they asked 'Would you like to meet us, Dr. Cathey?'

I first met Denise Crompton, her husband Bob, and their adult daughter Kelley, in 2006. The Cromptons and thirteen other families impacted by ML arrived in Greenwood, South Carolina to participate in the world's first "ML Clinic" held on the campus of the Greenwood Genetic Center. Parents will travel quite a distance to help their children. For ML, help does not yet include an effective treatment. Those initial contacts and first trips to South Carolina launched ongoing natural history studies of ML and related rare disorders. How can we treat a disease if we don't know much about the people who have the disease?

In this book Denise Crompton introduces us to 15 children from 13 families affected with ML. Although these patients have an uncommon disease, common themes emerge when their stories are collected together. Some of the medical issues are unique to ML, but many of the challenges are shared by any family facing diagnosis of a rare disorder. Parents will quickly relate to the worries, the hopes, the disappointments, and the countless appointments. Because rare diseases are in fact, rare, families may encounter doctors, nurses, and therapists who are unfamiliar with a diagnosis. The frustrations of navigating a very complicated medical system are only magnified. Parents are faced with educating not only themselves, but the rest of the world about a disease they had probably never heard of until their child received the diagnosis.

This book should be required reading for students in all fields of medicine. All of our interactions with patients, routine or otherwise, are important parts of their journey and opportunities for us to learn from them. I am truly appreciative of these patients and their parents, and the lessons they have taught me. These are lessons that are not learned from medical textbooks.

Sara S. Cathey, M.D.

Greenwood Genetic Center

N. Charleston, South Carolina

FOREWARD II

Rare diseases is an unexpected journey and impacts the affected individual, family and their community. I was introduced to Denise Crompton through the National MPS Society. Her daughter Kelley had ML III, like my daughter Jennifer. Denise was the very first parent that reached out to me in support. She was persistent though I was months in denial. Denise broke down the walls and showed me and my daughter how to live life with a rare disease.

Denise is not only a mentor to many families with Mucolipidosis but she is also a second mother to many. Her advice, presentations and availability to converse with families and physicians around the world has brought comfort to many while facing the most difficult decisions. Denise's inspiration was her beautiful daughter Kelley. Kelley has been an ongoing inspiration to those affected by Mucolipidosis because she lived a full life and was determined to enjoy each day.

Denise wrote Kelley's Journey over 11 years ago while Kelley was still alive. *Diagnosis: Rare Diseases* was written after Kelley's passing. It focuses on a group of families that struggle and still have learned to live a fulfilled life while maneuvering this uncommon journey from the day of diagnosis and beyond.

Denise has showed the world of rare diseases once again, that together we are strong! It is through these efforts that families will continue to unite with hope for a treatment and a cure. These will no longer be simply words for our children, but a reality. Denise and her family have always been humble servants – their footprint in the world of Mucolipidosis will forever light our paths.

Terri Klein, Mother to Jennifer, ML III

Development Director, *National MPS Society*

INTRODUCTION

According to the National Institutes of Health, a rare (or orphan) disease is generally considered to have a prevalence of fewer than 200,000 affected individuals in the United States. But collectively, rare diseases affect millions of Americans of all ages. These diseases are often serious and life altering, with many being life threatening or fatal. Because each rare disease affects a relatively small population, obtaining a diagnosis and finding treatments can be a daunting task for individuals and their families.

The National Organization for Rare Disorders (NORD) website provides the information that there are more than 6,000 rare disorders that, taken together, affect approximately 30 million Americans. And they provide reports on more than 1,200 diseases.

Those statistics mean little to the family learning that one of their members has one of these rare diseases. They often feel isolated and terribly alone. The reactions any individual might have to hearing a doctor give them a strange sounding term, as well as a prognosis that little is known about the rare disease, will vary. Usually shock, coupled with fear, is the first response. Whether the problem shows up before birth, soon after birth, or many years later, it is usually unexpected, and quickly leads to questions. Why did this happen? Did I do something wrong? Is someone else to blame? Is the affected going to die? Will the affected have a normal life? What kind of symptoms will they experience? What can be done about it? If immediate treatment is needed, families sometimes plunge into emotional turmoil, depending upon the seriousness of the situation.

While everyone has questions, not everyone is ready to hear answers immediately. We each have our own timetable for processing information. Reactions can vary from crawling back under the covers to amassing an army to fight the problem. Today, the internet has become a wonderful tool for people to connect with others and find information that was once found only in medical libraries. The connections that I have made through the internet have enlightened me to the myriad of problems others have had to deal with when confronting the diagnosis of a rare disease. I have come to admire the way in which I have seen some families rise to the occasion of their extraordinary challenges.

In 2003, my book *Kelley's Journey: Facing a Rare Disease With Courage* told the story of one family dealing with a rare disease since 1966. I wrote the book in order to reach out to others who were on the same path. At that time, our family had been in contact with only a few other families who shared the same diagnosis. Since then, we have been able to meet, and connect with, even more. This book focuses more on our collective experiences, in order to bring more awareness of the need for attention to rare diseases. Some of the original text of *Kelley's Journey* is included in the following pages. She survived for six more years after I finished that book. It is important for me to complete that story, including the interaction we were able to have with other families. If you have experienced the many twists and turns that the path a rare disease leads you down, you will find many similarities with the families whose stories are within. You will recognize the situations that are related, even though the particular doctors and hospitals are in different locations. Most of the health care workers that we have met have been humble, dedicated and caring individuals. However, we did have some negative experiences along the way. Since the situations, and the effects they had, are paramount, the names of particular doctors or hospitals are omitted. It is my hope that these stories will bring about awareness, which is the first step toward reaching the day when answers are found for all of those who are affected by rare diseases.

Denise Crompton

CHAPTER ONE

DISCOVERY

In 1966 my husband and I took a short vacation while my mother and my sister, Mary Lou, took care of our three young children. Shortly after our return, Mary Lou questioned us about the stiffness she noticed in the hands of our oldest daughter, Kelley. Since we weren't aware that anything was wrong with Kelley's hands, we called her from her play.

The three-year-old whirlwind ran into the living room and threw herself at me, laughing. I sank back into the chair, picking her up and laughing with her. What a fun kid!

"So, did Aunt Mary Lou teach you any new songs or games while we were away?" I asked.

"I dunno," she giggled while I looked at her fingers and glanced over at Bob.

"Come here, Kelley," he said, taking her from me. He sat down with her on his lap and began to look at her hands.

"Can you do this?" Bob asked, making a fist.

"Nope," she chirped, unconcerned.

"Let me see what you can do," he coaxed, wrapping his large hand around her little one. Her fingers simply would not bend all the way to make a complete fist.

"You can go play now," her daddy said, putting her down from his lap. She trotted off to resume her activities with her two-year-old sister, Susan.

"I'm shocked!" I exclaimed, after she left the room, "How did I miss that? I must be a terrible mother!"

"Well dear, you have had your hands full," my mother reassured. She was referring to the fact that, in addition to Kelley and Susan, we were proud parents of our nine-month-old son, David.

It was too late for us to get in to see the doctor that day, but our doctor's secretary found an opening on Monday. It would be a long weekend.

We called Kelley our little firecracker, because she was born on July 4, 1963, and displayed a vivacious energy that indeed sparkled! She didn't simply like something; she *loved* it. She never sulked or whimpered. She *hollered*.

Like most children, Kelley didn't like to take "no" for an answer. When she couldn't get what she wanted one way, I could count on her to try another. By the time she was three, the little darling knew enough to put her younger sister up to doing things she knew would not meet with my approval. I got after Susan numerous times before realizing she had become a scapegoat.

In general, my oldest daughter managed to get into, under, around and over more things than my imagination could. She always had another trick up her sleeve. I child-proofed our home before she was born, by putting anything potentially dangerous out of the way. However, I needed to watch her all the time, because she was very active.

Just when the problem with Kelley's hands started, I have no way of knowing. It had happened so gradually that those of us who were with her daily didn't see it.

A visit with our family doctor led him to making an appointment for us at a big city hospital, while he tried to reassure me by saying he didn't think we needed to be alarmed.

It was weeks before they could see us. Meanwhile, frightening thoughts went through my head, and I prayed. I have always prayed. I didn't wait for this to happen to start praying, but at this point, my prayers became intense. I asked God to help me to accept His will, and I fought back tears when I allowed myself to dwell on the things that could possibly be wrong. I learned that trying not to think about a problem was easier said than done, but I didn't speculate aloud. Bob is a private person who usually keeps his feelings to himself. I knew he was wrestling with this problem as much as I, and felt it would add to his burden if I revealed my apprehension. We waited in silence.

I was very busy in those days feeding, washing, and changing not one, but three babies, and I loved doing so. Other than Susan's allergy to milk, they were all healthy. I knew I was blessed to have them. They were a constant source of wonder and joy!

When our appointment day finally arrived, the first impression I received at the big hospital was that the kids with the sad faces belonged to parents with sad faces. I promptly made a decision that, no matter what we had to endure (and an inner voice told me we were in for a long haul), we would not become morose.

It's difficult to remember all the events of those first few years. Some were forgotten as quickly as possible, to retain sanity. Some I wanted to forget, but I couldn't. Fighting discouragement became a challenge.

It took an hour to drive from home to the big city. We were a one-car family in those days, so I was fortunate to have healthy, helpful parents. For every appointment, Grandpa brought Nana to the house to mind the two younger children, while he drove Kelley and me to the hospital. It wasn't possible for Bob to get away from work for the many visits. Some of the doctors cared only about her joints and drew bleak pictures about the future of a child who had, at that time, been diagnosed with Juvenile Rheumatoid Arthritis (JRA). They made remarks indicating that her condition would become worse; she might never be well enough to go to school; there was no way to know what to expect. They also advised us that we shouldn't allow her to become upset, because that could cause a flare-up.

Even the most cockeyed optimist can find it difficult to remain positive under such circumstances, but Kelley had a darling smile and a spirit which loved a good time. Grandpa and I put all of our efforts into making each trip as much fun as possible. We planned outings, discussing where we would stop to eat on the way home. Kelley always wore a special dress for the occasion, and we took along her favorite books and toys. The way Grandpa and I acted, one would think we were taking the child to a circus rather than a clinic.

We never used baby talk with our children. The best way for them to become articulate is for adults to speak properly to them. However, it appeared that some nurses and doctors never progressed beyond baby talk themselves. Kelley was a child with whom you could level, but if you talked down to her, she became insulted and wouldn't cooperate at all. There is one picture I still recall clearly. Two doctors were talking to me *about* her *in front of* her, as though she couldn't comprehend the language. They then turned to her using baby talk. Apparently she reasoned that if they wanted to play games, she would play games with them. It always brings a smile to my lips when I picture four-year-old Kelley, in her pretty white dress with the red trim, and her fair hair in bouncy ponytails. She was squatting under a table looking at two doctors, with a big grin on her face, as they were bending over trying to coax her out, using this strange language of baby talk.

The clinic trips were wearing at best. Grandpa never complained about the hours of waiting, but it was difficult to keep Kelley amused and in good spirits during those long hours. We met many different doctors, most of whom were quite impersonal. Occasionally we saw the same one twice, but mostly we experienced rounds of the same questions over and over again. My questions were often brushed off, giving me the impression that they didn't think I should be asking any. Never once were we seen at the appointed time. We endured long waits at the clinic, long waits at the lab, and long waits in x-ray.

Often someone wanted to take Kelley away from me for some lab work or examination. I didn't like it when someone said, "No, mother, you can't come."

I wanted to respect their positions, but I knew my little girl needed me. How could I submit her to having strange people examine her, or stick needles in her arm, and not be there to hold her hand? Maternal instinct was strong, so I learned to assert myself by saying, "If I don't go, she doesn't go, either."

Grandpa and I laughed about the 'busy' personnel at that clinic. We sat in the various waiting rooms, waiting our turn, and watching them perform their tasks. Rarely did we see anyone even walk quickly, yet we often heard them complain about how far behind they were, and how overworked they were.

My father was a great stabilizer. On the way home from our clinic visits, we always stopped to eat somewhere. That's when he would tell his stories. Never one to lecture, or to give unsolicited advice, he could get a point across just fine. He helped me enormously in coming to grips with the situation. After letting me vent my feelings, he would say, "That reminds me of the time that...."

We always ended up laughing. Fred Kelley never ceased seeing the humorous side of life. What a blessing to have such a father! In the Kelley family, we were never allowed to feel sorry for ourselves. Now, I was not allowed to feel sorry for myself, and I could not allow myself to feel sorry for my daughter, or to let her feel sorry for herself. Self-pity can breed self-destruction.

Kelley's hands seemed to be the most important problem in the beginning, although it soon became apparent she was experiencing limited range of motion in all of her joints. The immediate treatment included physical therapy for her joints and braces for her hands. These braces were very elaborate and difficult to adjust. It was also very difficult for Kelley to become adjusted to them. Although the brace maker was the kindest person we met there, Kelley did have some difficulty warming up to him. After all, he was the man who put those cumbersome things on her hands that interfered with her fun and games.

At first, Kelley wore the braces for a few hours at a time. Once she adjusted to them, she was expected to wear them during all her waking hours. They were constructed with wire, and little elastics were used for tension and flexibility. These elastics were tightened as her fingers started to bend more. We broke many types of elastic during that time! We were instructed to soak her hands in warm water a few times a day, and follow up with a dull Physical Therapy routine. Kelley did not like PT. I tried to find different ways to do it, in an attempt to make it more pleasant. By doing it right before lunch and right before supper, we had something pleasant to look forward to after finishing. We did it to music. We made up songs. We made up games. We invited other children to join the games. I prayed a lot, yet she still didn't like PT. She liked the braces even less. They limited her ability to do many of the things she wanted to do. I wondered how effective they really were.

When the experts at the hospital decided that Kelley's elbow joints were constricting too much, they made splints for her to wear to bed at night, which kept her arms straight out while she slept. The first night I put her to bed like this, she looked at me in panic, protesting, "I can't reach my thumb!"

I felt like a very mean mother. I didn't think it was fair to take away her thumb. For years she hadn't sucked her thumb during the day, but, when going to sleep, she found it to be a comfort, along with hugging her favorite stuffed animal. When I removed her splints in the morning, her arms resembled springs being set free as she grabbed for her beloved stuffed animal and put her thumb in her mouth to make up for lost time.

As soon as breakfast was done each morning, Kelley wore the braces until lunchtime. She soon learned to stretch out breakfast, lunch and dinner. Who could blame her? I didn't rush her. I never once liked putting those things on her hands. They said it was necessary to help her joints, but I wondered about her spirit.

In searching for answers about JRA, I frequently learned more by comparing notes with other mothers in the waiting room than from the doctors. We saw some children at the clinic with very severe cases. I wondered if that was what we should expect for Kelley. No one could answer that question. There appeared to be no reason for the onset, no true course for the disease to run, nor hope for a lasting remission.

In addition to the splints, braces and physical therapy, we were directed to give Kelley twenty-seven baby aspirins daily. We tried all the different brands and flavors before settling on the right one for her taste buds. If she had to chew nine of them with each meal, they might as well taste good.

While Kelley was somewhat belligerent at times about wearing the braces (she was known to hide them), and she wriggled out of those arm splints at night, she had spunk one had to admire. She decided not to let this problem stop her from having fun - period!

In the fifteen months following our first trip to the big city hospital in October 1966, I learned a good deal about JRA, including the fact that the doctors we saw there didn't know a lot about the disease. It seemed they had as many questions as I did.

One bitter cold January 1968 morning, we went to the city for what we expected to be a routine day, including waiting and lab work, waiting and PT, waiting and doctors. When the doctors examined Kelley, they asked if she had been experiencing pain in her legs, knees and hip joints. They seemed to be perplexed when she told them she had not. I told them she had been taking dancing lessons and doing as well as anyone else in the class. She also kept up with her playmates with no trouble.

Frowning, the chief clinic doctor said, "I don't like it. Her hip joints are not right. We'll admit her today to put her in traction."

I grimaced. I had been in traction a few times myself, and I was an adult at the time. What would it be like for this little four-year-old?

"Oh, it will only be for a few days," the doctor assured me.

Stunned, I went to my father in the waiting room and said, "There is no need for you to hang around here all day. They are going to admit her."

He frowned, and I shrugged my shoulders responding, "This is a renowned hospital. They must know what they are doing."

I made a list of the things she would need, so Bob could bring them in after work. I had a pounding headache by the end of that first day. I hadn't left Kelley to get myself anything to eat, or drink, as they kept telling me the doctor would be in to see me momentarily. That 'momentarily' lasted from noon until 9pm.

By 8pm, Bob arrived with his questions, and I had no answers for him. A nurse appeared to announce that visiting hours were over, and tell us we would have to leave. I emphatically informed her that I was not going anywhere until I saw the doctor assigned to Kelley, if I had to stay all night! The elusive doctor appeared in her room five minutes later, but we were disappointed by his vague answers to our questions.

When they admitted her, they had told us it would be only for a few days. After three days, Kelley started asking when she could go home. The answer was, "We'll see tomorrow."

When the next day arrived, the answer was again, "We'll see tomorrow."

That scene replayed day after day.

Kelley came down with a terrible cold, yet no one seemed to care. It was only her joints that interested them. We became disheartened. I thought that these people were supposed to know what they were doing, so I went along with their treatment. I didn't know what else to do. It *was* a very renowned hospital.

As the couple of days turned into a couple of weeks, I learned a new, more extensive physical therapy routine. We were instructed to create a traction set-up for her bed at home. Bob's expertise as an engineer made him well qualified to do a thorough job of preparing her bed to take the traction. It was necessary to put the bottom of her bed up on shock blocks, as had been done in the hospital. He had to make a special platform for her to lie on, in order achieve the right angle to counteract the weights. Otherwise, she would have slipped right out of the bottom of the bed. There were pulleys, ropes and clamps. I made bags to hold the sand.

After two weeks, we finally were able to take her home. They told us it was important that we not allow her to stand or sit. I was instructed to take her out of traction two times daily, carry her to the tub to soak in hot water for twenty minutes, and then go through the PT routine. Kelley didn't like PT, and I didn't like playing the heavy in making her do it, even though I believed it was for her benefit.

She still had to wear the braces on her hands in the day and splints on her arms at night. They directed us to return to the clinic in a month. At that time they would most likely discontinue the traction, put her in casts, and teach her to walk with crutches. A month is a long time to be tied down, so I started looking for help by reading up on caring for disabled children. I asked everyone I knew to visit. I did not want to make David and Susan spend all of their time in that room playing with Kelley, so I was thankful when they did so often, on their own.

Now, besides being a busy mother of three little children, one was bedridden. I had become so busy, that I made the mistake of neglecting to take care of myself. I became very sick from the flu. Bob had to stay home from work for a few days to take care of us. I learned then how important it was to take care of myself when Kelley was in the hospital. She needed me to be well when she returned home. I decided that was a lesson I should never forget.

The month passed, and we anxiously went to see the doctors, carrying Kelley. When the doctors did examine her, they told us to go home and add a pound of weight to the traction, so we could take her back the next month to get casts and crutches. We were very disappointed to realize that we had to face another whole month with her in traction.

Month after month we returned to the clinic with high hopes that this time they had meant what they had said. Month after month, we were met with the same shaking of heads. More and more I wondered if this was the right thing to do. More and more I questioned them, receiving only vague answers. While we waited for prayers to be answered, we kept following the doctors' orders.

One of the answers to our prayers was that my parents bought the house next door to us. Now I had some good reliable help close at hand, plus a generous supply of moral support.

Bob was able to take us to a clinic visit during his summer vacation. He learned firsthand why we always returned home exhausted. An engineer finds it difficult to witness waste of time and energy, as was apparent there. He felt the same frustrations as I did while trying to communicate with the doctors.

In September we went for yet *another* check-up, hoping for some good news. The doctors stood around shaking their heads while talking about attaching wires to Kelley's bones and putting her back in the hospital. With all she had been through already, and with the care (or lack thereof) I had already seen in that hospital, this idea frightened me! It would be too easy for her to get an infection. After nine months of traction, I decided that we had tried it their way long enough!

I continued to pray for guidance, and the next day, as I sat at my desk to work on a grocery list, I saw it! It had been there all along, but this day, I really 'saw it' for the first time. The pamphlet from the Arthritis Foundation made me realize where I could look for help. I wrote a letter to them, explaining what had happened, and what the doctors proposed to do next. I asked if they knew of any alternative treatments. I was feeling pretty desperate, but I knew somehow that this plea for help would be heard. I slept much more peacefully that night.

An answer to my cry for help arrived within two days. The medical director of the Arthritis Foundation suggested having Kelley admitted to a specialized hospital, only twenty miles from our home. In that way, he would be able to see her when he visited there weekly. Filled with expectancy, I shared my newfound hope with my parents. Grandpa offered to baby-sit while Nana and I went to investigate the hospital.

After we explained the reason for our visit to the receptionist, she cordially invited us to take a seat while she called the hospital's director. In a matter of minutes, he approached us, extended his hand, and introduced himself. We followed him to his office. Without rushing or interrupting me, he listened to my entire story. He explained to us that there were many children in residence who would probably spend their entire lives there. He didn't want Kelley to get the idea that she wouldn't get better and return home, so he planned to place her with another child who would also be leaving someday. I knew my prayers had been answered.

Shortly thereafter, we left her at that hospital knowing she was in capable hands. The building was well designed for the specific purpose of dealing with long-term illnesses. The type of nursing in this facility was very different from that in the big city hospital. Nurses became personally involved in caring for long-term patients. There was little sickness, a small amount of surgery, and a great demand for cheerfulness and encouragement. Out of necessity, the hospital staff became surrogate parents. Thus, they disciplined their charges as parents would.

The staff doctor in charge of Kelley's case shook his head upon initial examination. "What a shame to have kept her in traction for so long," he said, "Look how those muscles have atrophied. It will take a lot of work to get her walking again."

"Do you mean she *will* be able to walk again?" I excitedly asked, "Will it be okay for her to do so?"

"Oh, yes, she should, after enough physical therapy."

"I have been doing PT with her at home all along, but I was told not to let her walk. I did catch her at it a few times during the summer," I admitted, "When she was out in the yard with the other children at the picnic table during her hour's release, I looked out and saw her holding onto the table and making her way around by walking. I didn't have the heart to stop her and take away a moment of joy."

The doctor's eyes sparkled as he smiled saying, "We will have the specialist see her when he comes in on Tuesday. I am sure he will agree with me."

The specialist was the one who had answered my letter. He could not have been any kinder or more understanding, and to top that all off, he did know something about this strange disease. Every time the other doctors said they didn't know, I was under the impression they were talking for the entire medical profession. It turned out that there were some people in the medical field who did know a great deal. What a relief!

With no traction, splints or braces, Kelley thought this hospital was an okay place, even though they made her do physical therapy. It was many weeks before the muscles in her little legs were ready to try a few steps at a time. I was overwhelmed with gratitude and relief the day she actually walked again. As the tears ran down my cheeks, I noticed a few of the nurses had tears in their eyes, too. I started to have confidence again in the medical profession. That day Kelley put her arms around the doctor to give him a kiss when she was going home for a weekend, I knew her faith had been restored, too.

A few months after her admission to the 'answer to a prayer' hospital, Kelley was sent home. She returned on an outpatient basis twice a week for PT, and every few months to see the doctors. They still had many questions concerning Kelley's condition. She did not have the typical type of flare-ups known to happen with JRA, and the curve of her fingers was different from what they usually saw. They also said her blood work was not typical for that condition.

Thanksgiving 1968 was a very thankful one for our family. Our girl was home and, except for doing PT twice a day, we were a normal family with three preschool children. We did PT at home in the morning before school and again before supper. Twice a week, we went to the hospital for PT. They gave Kelley's hands a twenty-minute whirlpool bath before the exercises, which they kept updating. It was reassuring to have them educate me as to the reasons for each change. Some days Kelley really gave me a hard time about doing PT. She gave them a hard time at the hospital, too. It was difficult to impress upon such a young child the importance of keeping her joints moving, so they would not become any stiffer, and to impress upon her that she would be able to do more if she could loosen up those joints. Her resistance was to structured exercises. She wanted to be able to go off and play like the other kids did, and there were times we had a battle of wills.

When I was visiting my chiropractor for spinal adjustments, I told him about Kelley's problems. He suggested I bring her in sometime when I had an appointment, just so he could look at her. When I did take her, he asked to see her feet. I told him that it was her hands about which we were most concerned. He said, "I know, but I can tell a lot from looking at feet."

I took off her shoes, and Bob and I exchanged glances while the doctor examined first Kelley's feet and then her hands.

"She doesn't have Rheumatoid Arthritis," he said after a while, "I don't know what to call what she does have, but I know it is not JRA!"

I told him the doctors we were currently seeing had been questioning the diagnosis, also. He said he wouldn't be able to cure her, but he would be able to help her somewhat if I brought her along with me whenever I went there, and he wouldn't charge me for her. Thus began a beautiful patient-doctor relationship. Kelley grew to love the man. She enjoyed going to see him. He never stuck needles in her, he never gave her medicine, and he never talked down to her. She made cards and drew pictures for him, and he picked her up in his big bear arms to let her kiss him goodbye, each time she left. He saw her regularly, long after my back was better, at no charge - he insisted. He said she brightened his day, and that was all he needed to collect. I sent him quite a few patients, on the testimony of what he had done for us.

In January 1970, we sat in the conference room at the hospital with the doctors and nurses working on Kelley's case. Our specialist said, "I've shown Kelley's x-rays to a colleague at a large city hospital. I have been under the impression for some time that Kelley doesn't have JRA. This other doctor is in agreement with me. He would like to see her. He specializes in diagnosing questionable diseases. If she were my daughter, I would take her."

The idea of going back to the city hospital was not appealing. Yet I had confidence in this doctor. He had shown himself to be thoughtful, kind, thorough, and not given to drastic or dramatic treatment. It was a different hospital than the previous one, our new specialist was connected with it, and we really no longer had a diagnosis.

I agreed to go back to the city, and although I shook this doctor's hand when we left, I really wanted to give him a big hug. I let Kelley do that. At her age, she could get away with it, without embarrassing him.

Now, after all this time of learning about JRA and coming to grips with it, we found ourselves without a diagnosis. Little did we realize when we entered the waiting room of the new hospital that winter just how many more times we would cross that threshold. It was with anticipation that we first met the specialist who was going to have the challenge of coming up with a diagnosis. I wanted him to be a magician who could say, "Oh she has XYZ disease. A little lemonade twice a day will take care of that!"

Once again, we were asked a battery of questions. When did this show up? When did that happen? When did you first notice? What is happening now?

Kelley's new specialist said he was under the impression that Kelley had a birth defect. With the help of tests, x-rays, computers and other doctors, he would try to find out what the problem was. He ordered only x-rays and blood work that day, being careful to explain that he did not subject a child to tests just for the sake of compiling data, and that he would proceed slowly, testing only when he believed it was indicated.

This doctor appeared to be truly interested in what he was doing, as well as interested in his patient. He was very straightforward, and didn't talk down to us. He assured us that the diagnosis of JRA was a natural one for the previous doctors to have made, as the symptoms so closely paralleled the symptoms she had. At that point, it really didn't matter that Kelley had been misdiagnosed. We were now going forward looking to finding the real diagnosis.

As in the past, Grandpa took us to the city. It was wonderful to have his company. I didn't return home as tired from this visit as I had been a few years before. At least it looked as if things were being done correctly in this place. That gave me a good deal of hope for Kelley's future.

The testing started. As each evaluation came in, our new specialist discussed it with us. Kelley was shorter than average for her age, and weighed less, too. We already knew that, but we had not known before then that her x-rays showed all of her bones to be different from the norm. New light was being shed with each visit. When the specialist told us that he was sure that Kelley never did have JRA, my first question was, "Can we take her off the aspirin?"

I was warned to do so slowly, which I did. She experienced withdrawal symptoms, nevertheless.

Around this time, a problem with Kelley's eyes turned out to be a caused by muscle weakness. It was nicely corrected with the right glasses. She was happy to wear the glasses, as they put the world back together and she stopped seeing double.

A small piece of skin was taken from Kelley's arm, so they could grow it in a lab for a few months, and then conduct some testing on it, with the expectation that they might be able to help identify her problem. A presentation of her case was also scheduled to take place at a conference in that hospital in May. The approach the doctors were using then was to rule out the condition they thought might be Mucopolysaccharidosis VI (MPS VI). When they gave me that name, I looked for information at a local hospital library and found that, with this particular disease, there was a chance she would not survive beyond the age of twenty. However, her symptoms should have been progressing more rapidly, so I decided to adopt the attitude that she would live. The temptation to overindulge her, if she was going to die young, would have been great. But if she lived longer, she could have ended up being a spoiled brat. I reasoned that, since no one wanted to associate with a spoiled brat, spoiling her was not a good idea.

Neuropsychological testing was done on Kelley in April 1970. It took a full day to complete. The evaluation showed definite motor difficulties. They found Kelley to be cooperative, and approximately one year behind her chronological level at school. They questioned the possibility of over-protection. The doctor who asked me about this was trying to be tactful.

"Of course she has been overprotected," I answered, "We went seeking medical advice for her when she was three years old, and since then, up until a few months ago, we've been told she had JRA. She needed frequent rest, and as little emotional turmoil as possible, because emotional reactions could cause flare-ups. We were, in essence, advised to handle her with kid gloves. Yes, I would say she has been overprotected!"

The doctor laughed and said, "Well, she doesn't have JRA now, so there is no need to overprotect anymore, but I do suggest you make the change slowly."

It was reassuring to be receiving some concrete help for the whole child. I was grateful to have this pointed out to me.

When May arrived, Bob was able to go with us when Kelley was presented at a large conference of world-renowned physicians. Her doctor was hopeful that some other doctor would be able to shed more light on the condition. We felt fortunate to be able to tap into their knowledge and experience in this way. After the doctors asked many questions, they told us their consensus was that we were on the right track. Some of her clinical work up didn't fall all the way into the pattern of MPS VI, but they had not yet been able to completely rule it out. The decision was made to follow the path on which we had started.

When we visited Kelley's doctor that September, he said he had come to the conclusion that her problem was genetic, and that both Bob and I each carried a very rare gene. It was explained that statistically, one in four of our children would have the problem, but these things didn't always follow the statistical probabilities. I did not want to imagine what it would have been like to have two or three of them in traction at the same time.

The following winter, when the doctors noticed I was wearing maternity clothes, they suggested that I take the new baby to see them the next time they saw Kelley, so they could check her out.

Peggy was born on May 18, 1971. By that time, seven-year-old Susan was much taller than eight-year-old Kelley, and six-year-old David was a tall as Kelley, and ready to pass her in height. Kelley's first remark when she saw Peggy was, "Well, it will take a while for *her* to pass me by!"

We took Peggy to the specialists for a checkup when she was six weeks old, and were relieved when they told us that they were sure she did not have the same disorder as Kelley.

Kelley was more cooperative with PT (or she had resigned herself to it), by this time. She was able to do a lot more for herself, even though the joints in her hands were still very tight. However, it was an unexpected development that she also had a noticeable improvement with mobility. No one in the medical profession knew why. I mentioned Kelley's visits with the chiropractor to the doctors in the city, but they refused to believe that could have been the reason. I believed differently and kept taking her regularly to the chiropractor.

In the fall of 1972 we received the shocking news that the company, for which Bob worked, planned to close the plant where he was employed. The area in which we lived offered little opportunity for an industrial engineer, so when they offered him a transfer, he accepted. Reluctantly, we made the decision to sell our house and find one closer to his new work location.

Shortly after we moved, there was another conference to attend in the city. This one, put on by the March of Dimes for the National Birth Defects Foundation, was much larger than the one we had attended previously. Bob was not able to join us this time, and I no longer had Grandpa next door to take Kelley and me, or Mom to baby-sit for the older children. I had to find a baby-sitter quickly, and find my way into the city, which was still an hour away, but from a different direction.

The presentation was very interesting. Kelley and I sat up on a stage, in front of the audience. Her x-rays were shown on a large screen, while a doctor pointed out the abnormalities with her bones and joints, and ran off a list of various lab reports from blood samples. All of the other children presented that day were babies, but Kelley was almost ten, and the doctors had many questions for us. One of our doctors took Kelley into the audience for the others to see her up close. Later, a doctor, who said he had seen hands like hers before, approached us. He said he would be able to suggest some more tests. I had not expected anyone to come up with an instant cure that very day, but I had hoped for a little more than a recommendation of more tests. Nevertheless, I knew we had been exposed to some of the world's most informed physicians, so there was nothing more that could be done.

In the fall Kelley was scheduled for another day of neuropsychological testing in the city. When we received the results they assisted us in giving the school personnel a better idea of what they could realistically expect from her.

That was also the fall when my Dad, who had been quite healthy most of his life, was admitted to the hospital. By Christmas, it became obvious that he was suffering from congestive heart failure, and we were probably going to be celebrating Christmas with him for the last time. It all happened quite quickly. At the age of seventy-five, he was still ice-skating - indeed, teaching others to skate, and before his seventy-fifth year was over, he was gone. We were fortunate, however, that we all had enough time to prepare, and to say good-bye.

It has turned out, though, that in many ways, he is not gone. He was such a wonderful man, who shared so much of himself with us that his spirit has stayed with us all. A short time after his death, I was involved in a minor traffic accident. I felt rather annoyed at the man who hit my little car with his big one. His did not have a scratch. Eight-year-old David helped me put it all in perspective when he asked, "What would Grandpa say?"

Of course, Grandpa would have laughed, and said, "As long as no one got hurt, it's okay."

People always said that you knew when Fred Kelley was around, because you could hear him laughing. Could there be a better legacy? I've often thought that he has been there with us as we continued along the journey to find the right people to help Kelley.

CHAPTER TWO

RECEIVING THE DIAGNOSIS

Kelley was ten years old when all of the new testing had finally been completed. We were told that Mucopolysaccharidosis (MPS) had been ruled out, and that Kelley's condition was called Pseudo-Hurler Polydystrophy; later also termed Mucolipidosis III (ML III). This meant she probably would not die before the age of twenty, but there was not enough information on the condition for them to give us any prognosis. I asked the doctor about access to the medical library of the school affiliated with the hospital. He accommodated me with a smile, saying that I would only find one paragraph about the condition there. He was right, of course. In that paragraph, I read that some doctors had tried giving the enzymes they thought were missing, with no positive results. They drew the conclusion that there was much more to the disease than they knew at that time.

So, on the day that we received the diagnosis that our daughter had an ultra-rare disease, for which there was no cure or treatment, we learned that there really was no prognosis. All we could do would be to treat whatever problems she had as they appeared. We really had no idea what we were facing, so we were left feeling very unsettled.

I have often heard people say that it is natural to ask, "Why me?" I have never felt that way. I have yet to find a person who has not had problems, so why not me?

Why Kelley? Well, why not Kelley? I didn't like the situation. However, because my trust is in God, I believed that, no matter what we were to discover next, the ability to handle it would be given to us. I was thankful that we were not too far from the big city, and that we had a family willing and able to help us.

During those early years, as we were coming to grips with the fact that our daughter had a rare disease for which there was no cure, we found ourselves trying to explain the situation many times. When we were at the hospital that was affiliated with a medical school, our situation was pretty well understood, even if the disease wasn't. The staff there saw numerous people with rare diseases, so there was no need to explain what it meant to have something that was misunderstood by the rest of the population. However, we did experience considerable frustration with understanding and acceptance from the general public. Besides answering the questions from those who really cared, we endured others who made judgments about the way we handled our daughter. Well-meaning folks provided us with wonderful (and not-so-wonderful) ideas as to where we should seek help and what we should do next. There was some loneliness associated with dealing with such a rare condition. We knew no one else who was coping with the same disease, nor had any way to find someone else with it. During those initial years, the internet did not exist. It was not until many years later that we found the National Organization for Rare Disorders (NORD). That group helped us to connect with the Mucopolysaccharidosis (MPS) Society. Our initial contacts were through the telephone and mail. Eventually, the internet helped us correspond with others through emails.

We first found families of those affected with the various MPS conditions that are similar to ML III. Many of the children had hands that were shaped like Kelley's were, and they all had numerous physical problems caused by an accumulation of unwanted materials in the body. After years with the MPS Society, we started to connect with more ML patients in the United States and Canada. Once the International Society for Mannosidosis and Related Diseases (ISMRD) was established, we were able to find others worldwide. We also found that this particular disease has no boundaries. It strikes across all nationalities and races, with no discrimination at all.

In most cases, mothers have been the front-line advocates for their children in dealing with the doctors and the schools, so they have been the parents with whom I have had the most extensive contact. A dozen of these mothers have agreed to share some of their experiences while dealing with basically the same disease as Kelley had, with variations of severity and complexities. Their experiences are not unlike the frustrations that most people go through when embarking upon a hunt for a diagnosis and treatment for any rare disease.

The initial diagnosis of a rare disease was overwhelming to most of us, even though we were still unaware of the roller coaster ride we would be taking with our children. Many of our experiences could not have been anticipated, since each individual has their own course with their own symptoms. For most, the restriction to the normal activities of daily living caused by the disorder is only a partial picture. Recurring pain, often-relentless pain, becomes a challenge to the patients and doctors alike. The disease encompasses our lives, and family members are frustrated when trying to help. For many, hearing the diagnosis is such a shock that nothing is heard afterward. Within a week, they will have many questions that the doctors might have answered initially, but they simply didn't hear those answers. In our own case, there were no answers at the time of Kelley's diagnosis. We were to learn about the many symptoms as the years progressed.

Autumn Tobey

05/17/2012

Despite the fact that Autumn Tobey was born thirteen years after Kelley, some of their family's story was similar to ours, in that it took them ten years of working their way through eight doctors while hunting to find the reason for Autumn's numerous physical problems, primarily in her bones and joints.

Pam and David Tobey of Arkansas are the parents of two daughters. The youngest, Autumn, born in 1976, appeared to be a healthy little girl. Although she was small for her age and didn't walk or crawl as quickly as her older sister, Michelle, had done, Pam assumed it was simply because all children are different in development. In fact, it wasn't until Autumn was seven years old that her initial symptom showed up. While Pam was trimming the child's fingernails one day, it appeared to her that Autumn was purposely holding her fingers in a bent position. Pam asked her to relax and loosen her fingers, but the little girl said that she couldn't. That was when Pam first knew something was not right, and she hastened to make an appointment with their family doctor.

He sent them on to a pediatrician, who in turn sent them to a rheumatologist who only prescribed aspirin and followed Autumn for a few years. Even though the blood work never indicated the diagnosis of Juvenile Rheumatoid Arthritis (JRA), there appeared to be nothing else that could be causing her problems. During this time little Autumn started to become withdrawn, because she was scared and didn't understand why she was experiencing so much pain.

Next she saw another pediatrician, who realized that she needed a spinal fusion. The orthopedic surgeon followed that surgery by a carpal tunnel release in both hands, a tendon release in her right hand and a synovectomy, to remove inflamed tissue that was causing pain in her left ankle. A consult with another rheumatologist prompted a visit to an ophthalmologist, since JRA children were prone to eye infections. That doctor discovered that Autumn had cloudy corneas. He said she didn't really have JRA, but her problems were being caused by a rare condition. Thus, he recommended a consult with a geneticist, where she went through more testing, including much more lab work, while they looked at the MPS conditions, which were ruled out, as they finally made the diagnosis of ML III.

Although that news was puzzling and shocking, Pam's first thought was that now Autumn would get the help she needed. However, it was impossible to ignore the sickness she felt in the pit of her stomach... a feeling that didn't go away for a very long time, since the diagnosis left the family feeling very isolated and alone. And Pam was brokenhearted to realize that her daughter might never have a life like other girls... no boyfriends, no wedding, no husband and no children, and possibly no job.

When Pam and I made contact with each other in 2001, I understood her expression of relief in being able to talk to someone else who had "walked in the same shoes."

To the untrained eye, Autumn has no immediately obvious physical limitations, since she has what is now considered to be one of the milder cases of ML III. Measuring 4'11" in height, she is taller than most of those with ML III. Nevertheless, her life has not been easy. Her bones have disintegrated as a result of the disease, and she has had many extensive and painful operations in order for her to continue to function.

One advantage that the Tobey family had by entering the medical scene thirteen years later than Kelley did, was that the hospital workers had started to adopt the attitude that families were helpful in assisting with the care of the patients, and bringing about positive outcomes. I like to think that those of us who put up a fuss numerous times through the years really were heard, and we helped to bring about those changes.

Allison Dennis

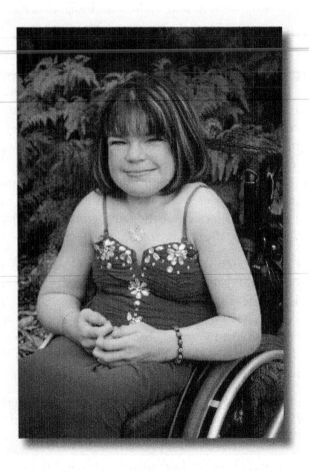

In Australia, Trish and Richard Dennis were told that their daughter Allison, born in 1983, was just fine. But from the beginning feeding the baby had been difficult, with Trish having to place her thumb under Allison's chin to help her suckle. In fact, one feeding would take so long that it just about ran into the next feeding. When they started questioning her growth and development, Allison's parents were told they were being paranoid. They were assured that all babies do things at different ages, so there really was nothing for them to worry about. Still, it troubled them that all of Alli's milestones were being met later than expected. She did not sit up until she was eleven-months-old, crawling started at fourteen-months, and her first step was at twenty-months. So, by the time she reached the age of three and was finally diagnosed as having a developmental disability, her parents were not surprised. There didn't seem to be any reason for this diagnosis at the time. However, when Alli was four, her hands started showing signs of clawing, just as Kelley's had done. Their general practitioner sent her to see a physical therapist, since she wasn't able to use her hands in the usual way. The physical therapist, noting that Alli's hands might be indicative of a more serious problem, believed it was important to refer them to a pediatrician. That doctor had seen hands like hers previously in a child who had MPS, so he thought Alli might have that condition. He sent them to see a geneticist who, in turn, took a small piece of Alli's skin for a biopsy. At that point, Trish tried to do some research on MPS, only to discover there wasn't much information in the library of their local hospital; the internet was not yet available. During the long and difficult eight months they waited for the results of the testing, Trish felt confused and frightened, hoping they had made a mistake and wanting only the best for her daughter. Richard was so upset by this turn of events that he couldn't even discuss it. Alli was six years old when Trish and Richard learned she had the rare disease of ML III.

Although the confirmation of knowing there was a medical reason for all of Alli's problems provided them a feeling of relief, the diagnosis of ML III was overwhelming. They were told that a leaking enzyme caused the condition and Trish hoped that the enzyme could be injected, and her daughter would recover. But she was disappointed to learn that her hoped-for treatment was not to be, since the condition is far too complicated and there was no known treatment or cure.

Richard blamed himself, which is not an unusual reaction to such news. Some people feel a need to place blame. After the stress drove him into a depressive breakdown, Trish reacted by changing her personal plans for further education in order to focus on her family's needs. She never wanted to be in the position of thinking there was something more she could have done for her daughter, so she employed many prayers as she put all of her focus on caring for her family.

A bit of an ironic twist to their story is that sometime later, after the birth of their son, Nathan, a nurse at the baby health center did tell Trish that some of the staff there had thought that, in Allison's early years, she clearly had some problems that might point to Cerebral Palsy. Yet they had never told that to the Dennis family, despite the fact that they had questioned the baby's development on numerous occasions.

Hayden and Sarah Noble

Jenny and Paul Noble in New Zealand know all too well about the many complications caused by ML III. Their son, Hayden, born in 1981 and daughter, Sarah, born in 1986 are both affected. As it has been with other families dealing with rare diseases, their initial journey was one of loneliness, late diagnosis, shock, despair, and wondering what to do next. Hayden was six years old when he was diagnosed, and Sarah was two. Being born prematurely, Sarah had numerous problems that were handled in her early years. Once they moved past those issues, the Nobles believed they had a normal and healthy family. Yet, only a short time later they were informed that Hayden was not keeping up with his peers in growth and learning ability, so he needed a physiological assessment. It was disturbing to learn that he was two years behind a normal five-year-old. They enrolled him in a special education kindergarten with one-on-one help, where the teachers noticed he had trouble keeping up with the other children on the playground. That led the Nobles to seek the help of a specialist in a pediatric clinic.

After Hayden was put through examinations and testing, his young parents were told that he had ML III, and since it was a genetic condition, their other children should be tested. They were relieved that their son David, born in 1983, was spared the condition but distraught to learn that Sarah tested positive.

Jenny and Paul were given a two-line explanation of what ML III was and advised to go home and get on with their own lives, since their children would not know what a normal life would be like. Jenny felt like she had been flung into a black hole, with no idea of how to get out. She was saddened to think that the young boy who always had a bat and ball in his hands would never be able to play a sport. Paul had to come to terms with the realization that he would never be able to walk his pretty little daughter down the aisle. With no idea of what really was in store for their children, the young couple desperately started to search for answers, while at the same time starting to plan for a future that would include taking care of their children for the rest of their lives. They were very disappointed to learn there might only be 60 children in the world with the same condition. Since they were unable to find any other families in New Zealand with ML III, they focused on Australia as the next logical place to look. In time they were relieved to be able to connect with others in Australia through the MPS Society, which they considered to be a lifesaver as they faced a very uncertain future for their children.

Hudson and Samuel Anthony

Back in the United States, Liz Anthony agreed with a friend who once described the three boys born to her and her husband Tom, "the best boys in the whole world." Hudson (Huddy) was born in July 1989, and Samuel (Sammy) was born in April 1991. Huddy displayed some developmental delays, autistic-like behaviors and some physical features that caused concern. Before the family found a geneticist who realized that testing for a diagnosis was necessary, Huddy was seen by a neurologist and two geneticists who could not diagnose him correctly. The boy was about three years old when the geneticist suggested that his condition was Hunter's or Hurler's. In looking over the booklets that she gave them from the MPS Society, Liz saw the picture of the little blond-haired boy on the cover of the booklet for I-Cell, also known as Mucolipidosis type II (ML II), and immediately knew that was the disease Huddy had. The boy's features were almost identical to Huddy's. It was fully a year prior to receiving the actual diagnosis for ML II, which was confirmed after the new geneticist sent a blood sample from Huddy to a lab for testing.

Sammy's cognitive and social skills were higher, so his condition was less obvious. Yet, they had blood drawn from Sammy for the same test. Tom was of the opinion that Sammy was normal and did not see the point of that testing. However, Liz could see that he had developmental delays. The therapists treating Huddy had watched for symptoms in Sammy from the time that he was three-months-old; they also saw delays and similarities.

Huddy was four years old and Sammy was two when the diagnosis was confirmed in July of 1993, and Tom and Liz were called in to the pediatrician's office to discuss the results of the testing. He confirmed that the test showed ML II in Sammy as well as Huddy. They were told that a bone marrow transplant was extremely risky, had only been done for ML II once (at that time) and both boys were too old to be considered candidates. The prognosis was that each boy could expect a lifespan of eight to ten years.

Tom, being a realist and seeing all the consequences of the boys' disease leading up to early death, was crushed by the news. Liz's reaction to the diagnosis was one of acceptance. She had already realized that this was probably the disease her boys both had, and had come to terms with that fact. During the prior years of therapy without a diagnosis, she had simply loved them and taken them to therapy to treat their symptoms. It did not matter to her that the symptoms had no name/diagnosis. What did matter was that the boys needed help and she loved them. Each boy had a distinct and gentle, engaging personality. Liz could not imagine how they could be any more perfect! Although these parents have different perspectives and feelings, they both share a strong faith in God and the Bible. They were well equipped to start upon a routine of therapy which, for a while, included four visits per week at two locations, medical specialist appointments, a rigorous home therapy program, respiratory therapy treatment, urgent-care trips, sleep studies, pneumonia, plus normal childhood illnesses. Their son, Aaron, who was born later, is unaffected.

Callie Nagle

About fifteen years after we moved away from the town of Chelmsford in Massachusetts to New Hampshire, we learned about Callie Nagle, daughter of Debbie and Richard, who was born in 1990. Since so many of the people with whom we have been in contact live many miles away, we were quite surprised to discover this family. Before Callie was five years old, she had needed two inguinal hernia repairs. Both of her parents were present for all of her appointments, as well as her surgery, pre-op and post-op. However, Richard was not able to accompany Debbie and Callie when they met with an orthopedic doctor to discuss the problem Callie was having when her fingers kept locking on her, and she was having trouble raising her arms over her head. The shock she received that day stays alive in Debbie's memory, as though it was just yesterday. After the doctor took a number of x-rays, he returned to announce, "Callie has abnormalities in all of her joints."

As happens to many people when being given such news, Debbie says that she didn't remember much of what he said after that. She did take away the fact that the doctor had no idea what was causing these abnormalities and could only refer them out for more testing. She knew that she couldn't ask any questions because he apparently had no answers. Shaken and frightened, she stood at the secretary's desk to schedule another appointment, desperately trying to fight off tears. Since little Callie was standing there with her, she called upon all of her inner power to keep from falling apart. She could only think of cancer because that was one of her worst fears. In retrospect, Debbie says that she compared the unknown diagnosis that day to her worst fear of a diagnosis of cancer. This was the beginning of a pattern of thinking it could always be worse.

They were referred to the Genetics Department, where a simple blood test was done, after which they were told that Callie was diagnosed with ML III. At that time she was five years old and the doctor they saw was unable to recommend anyone they could see who had experience with ML III. They found the whole ordeal to be truly agonizing, with difficulty in obtaining information about the disease, and more difficulty in finding doctors who were familiar with the disease. Equally frightening was the fact that there was no way to know how she would be affected, since each child is affected differently. As time passed, the Nagle family learned it was best for them to roll with the punches and handle each symptom and complication as they come along.

Jennifer Klein

Most children born with ML III appear to be completely healthy and normal for their first few years. Jennifer Klein, the youngest daughter of Terri and Walter, was born in Ann Arbor Michigan in 1992. She was a happy little girl leading a normal life, being active in gymnastics, dancing, soccer, skating and more, until she was in the second grade. It was then that an Occupational Therapist at her school noted a kyphosis (severe curvature) in her spine. The Klein family was shocked. Jenny had never complained of problems. Since she had regular check-ups with her pediatrician, her parents wondered why it had not been picked up previously.

Suddenly, this life-changing event plunged the family into a direction they never expected. For the following eighteen months, they searched for a diagnosis. They were sent to one doctor after another, exploring the possibility of one diagnosis after another, realizing along the way that Jennifer was beginning to experience difficulty completing common physical tasks. She started saying that she was in pain, even though she had never been one to complain. It was alarming that, once Jenny's kyphosis was discovered, progression was rapid. She underwent tests and more tests, ruling out a number of possible conditions, before they were finally told in 2000 that Jenny had the "Super Orphan Disease" of Mucolipidosis III. They were frightened to discover that, at that time, there were only three other known families in the United States who shared the same diagnosis.

Terri recalls going through an arduous journey of shock, denial, grief and awareness, after feeling like a lost lamb looking for her daughter's herd and looking for ways in which she could be of most help to her daughter, being concerned about both her physical and emotional health. Once she found the MPS Society, she joined, hoping to find others who were facing the same situations, knowing that we can all learn from each other. In 2001, she was able to start connecting with others through the internet, and learning that her family was no longer alone as they faced the days ahead.

Terri learned to juggle her schedule as a real estate broker because as Jenny grew, her physical problems also grew, leading to a number of major surgeries. Jenny also had many frightening battles with pneumonia, from which it was feared that she might not recover. And, as with the other families, they have turned to prayer to see them through the rough times. All of the parents we have met through this journey have told us that they gain strength by asking God for help. And as our children have gained in age, but not in height, we have found it necessary to handle the anguish of watching them grapple with the noticeable difference from their peers, as friends are able to grow taller and accomplish the physical tasks that their condition prohibits.

Andre Andrews

Andre Andrews was born in 1994 in Washington DC. His mother, Jane, says that her son was, for all appearances, a healthy, normal child until around the time he reached the age of two. His first symptoms showed up when he started having trouble walking. Jane noticed that his legs were becoming bowed. She pointed it out to Andre's pediatrician who ordered X-rays, and referred them to physical therapy. The physical medicine doctor said that he had never seen anything like the symptoms presented by Andre, so he sent them to see genetic specialist. Then it felt to them like "all hell broke loose." Andre was started in programs of physical therapy, occupational therapy, and speech therapy while they were searching for the reasons for his problems.

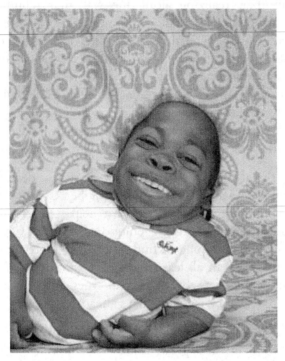

Shortly before Andre's third birthday, his parents were given the shocking news that his diagnosis was ML II (I-Cell). They felt devastated when they were advised go home and shower him with love, while taking plenty of pictures. At that time, the doctors believed that most ML II children didn't live past their third birthday. Little did they know that Andre was about to prove them wrong; as Jane decided to do everything she could to make the best life possible for her son. At first she felt like she was counting down his days, but as time went by, she adopted the philosophy of living life to the fullest, since no one can know what tomorrow will bring. She believes in appreciating the little things, realizing that each child is different and has different gifts that they give us. Being happy that her son can tell her that he loves her, Jane has chosen to ignore the "three year rule." Like so many of the heroic mothers I have met, Jane will do everything within her power to help her son, while always remembering that there is a Higher Power that knows all and will allow Andre to live a full and happy life.

Joey Nagy

Joey Nagy was born to Linda and Frank in 1995 in Illinois. He was only three-days-old when the doctor told his parents that their baby had some troubling physical problems, after he noticed contractures in Joey's hips and other joints. For the first six months of his life, Joey received physical therapy and splinting for what was called Bilateral Hip Dysplasia and Arthrogryposis. To their relief, the treatment produced positive results. All appeared well until the following year, when it was discovered that Joey had a mild heart murmur and his head circumference was well below the normal range. The next specialists they saw were certain that Joey had MPS when they first examined him. Yet, the results of the blood work testing ruled out that diagnosis, and the real diagnosis remained a mystery.

Joey was only three years old when they felt it was necessary for him to have reconstructive bilateral hip surgery to reposition his hips. Since he also displayed developmental delays across the board, he was placed in an early intervention program for speech therapy. When Joey was six, there was still no real diagnosis for his problems. A surgeon performed hand surgery in an attempt to straighten the curvatures of his fingers on one hand, but was baffled as to why the surgery was not successful. Thus he didn't attempt to try again on the other hand. As Joey grew, the problems with his hands grew. He struggled with the fine motor skills required to manipulate buttons, snaps and zippers. Finally, when Joey was nine years old, the hand surgeon referred them to a Genetic specialist who decided to test for ML III and the true diagnosis was made.

After initially being relieved to have a name for her son's problems, Linda felt devastated, frightened and isolated due to the fact that she was told that it was an ultra-orphan disease. The anxiety she felt for her son's future was compounded by the fact that they had received the diagnosis by a phone message confirming the tentative diagnosis, and warning them not to have any more children, since there was no treatment and no hope of a cure. It angered Linda that they were given the information in that way and that the medical community had not taken into consideration the emotional devastation felt by a family receiving such a diagnosis. She felt they should have received some counseling to help them deal with the anxiety and stress.

Since that did not happen, Linda went searching for information on the internet and was relieved to find the MPS Society and ISMRD. She quickly became very involved with other parents with whom she could share and from whom she could learn, as she was determined to do all she could to help her son deal with his physical problems.

Anna James

Jackie and Bret James had a four-year-old son, Peter, when they became proud parents of their healthy daughter Anna in 1995 in St. Louis, Missouri. Although she sat up at the age of six months, as time went by Anna didn't crawl or pull herself up to a standing position. The doctors brushed off Jackie's concerns, telling her that since Anna was a second child she was probably being lazy. However, she was almost two years old before she started walking and talking. Once the doctors noticed that Anna was a little pigeon-toed, they ordered shoe braces and referred her to physical therapy. The x-rays they took at that time looked normal to them.

When Anna was three, she started taking ballet lessons. About a year later, the ballet teacher expressed concerns about Anna not being able to raise her hands over her head properly. Jackie had noticed that her elbows and fingers were bent, and had previously asked the pediatricians about it. He had not seemed to be too concerned but now that the ballet teacher had also noticed, Jackie took Anna back to the pediatrician's office. Although he was unavailable, the nurse who saw Anna immediately sent them to a diagnostic specialist, where they spent an entire day of testing.

The following morning Jackie was called at work and given the alarming news that the doctors wanted to meet with both her and Bret on that very day. As it turned out, they met that evening with both the diagnosis specialist and Anna's pediatrician at the pediatrician's office after hours, so it was eerily quiet. In the way that time feels like it is being suspended, it seemed to take forever for the doctors to say they believed Anna had one of the MPS conditions. The advice given to her parents was that they should not look it up on the internet, as Anna would not live to see her tenth birthday and the information on the internet would prove to be too distressing. This devastating news was followed by visits to numerous departments at the hospital, including neurology and oncology. It was the oncologist who said he didn't think Anna had MPS.

Two months later, they finally met with the geneticist, who confused Jackie by talking about Mucolipidosis. She told him that Anna had been diagnosed with MPS, to which he replied that he was horrified she had been given that diagnosis. He said that Anna had been tested for ML II, but in his opinion, after clinical evaluation, he believed it to be ML III. At that point, Anna's parents were relieved because the fact that she would have a longer life meant everything to them, after having been given the previous diagnosis.

Thus, they embarked upon on their journey to find information about the condition, as well as how to deal with it.

Spencer Gates

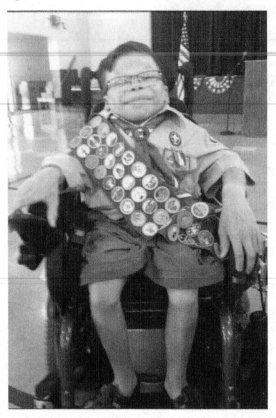

At the time Spencer Gates was born in 1997 in California, the doctors told his parents, Andrea and Kevin, that they had found calcium deposits in the placenta, which indicated that their baby had a storage disease. So, within his first few days of Spencer's life, they knew that testing would be required to discover what caused the deposits. However, the most common diseases were ruled out after the standard tests were concluded and his condition remained a mystery.

Still lacking a diagnosis, when Spencer was about six months old, they were advised he needed surgery to repair a hernia. As it happened, the same pathologist who had discovered the calcium deposits saw Spencer's name on the surgery list. He requested permission to do a biopsy while Spencer was under anesthesia, so he could further consult with doctors in Philadelphia. He received the official results of ML III when Spencer was almost ten months old.

It was six o'clock on Christmas Eve when Andrea received a most startling and upsetting phone call. She was told that they had determined the disease was ML III and that she and her husband should not make any plans for future children until they met with the geneticists, which should be as soon as possible. She was in shock but knew it had to be serious if they were advising against decisions to have more children. Yet, she felt some relief now that they finally had a name for his disease, since for many months they had been telling friends and relatives that he had a genetic disease. Every time she had to tell someone about the fact that he had an unknown disease, Andrea felt it was hard to explain. She usually ended up feeling helpless, embarrassed and downright dumb. Now that they finally had a diagnosis, at least she could start to learn about the disease and what they could do about it. While there was some relief there was also a profound sadness, as she was heartbroken for her son not knowing what his future held.

As advised, Andrea and Kevin met regularly with the genetic doctor. In time they started to realize that everything he was telling them was what they had already read by doing their own independent research. It became apparent to them that they and the doctor were both getting their information from the same internet sources, and they reasoned they probably knew just about as much as he did about ML. That was just the start of their very long road of learning while caring for Spencer and becoming connected with others on the same path. It is also a relief to his parents that Spencer's sister, Sydney, is unaffected.

Sergio Cardenas

Maria Elena and Gustavo were living in Venezuela when Sergio Cardenas was born in 2000. His mother knew from the start that something was wrong. It was a fear she had harbored since having a dream during her pregnancy. Her doctor conducted some tests to reassure her that she would have a healthy baby. However, she had two older children, and she felt a difference with this baby.

When Sergio was very young, Maria Elena started to take him to different doctors in various cities of Venezuela, looking for someone who could give them a diagnosis. However, all they did was diagnose Maria Elena as a mother who never overcame postpartum depression. She knew that was not true. Even though she could barely speak English, in 2002 Maria Elena and her sister traveled to the United States with her son, where they found a doctor who was a specialist in genetic bone diseases. Believing that Sergio had either MPS or ML, he sent them to see a geneticist, who then referred them to more specialists to start the process of testing which included X-Rays, MRI, CT scans, blood work, urine test, skin biopsy, echograms, breathing and sleep studies. When the testing was complete, the genetic specialists and a translator told them that MPS had been ruled out, and they were almost sure that Sergio had ML, but they couldn't determine if he had ML II or ML III. The diagnosis was subsequently described as ML II/III.

Maria Elena asked the doctor to give her all of the information in writing, so she could go to the computer and look for the information in her own language. She didn't show her emotions to the doctors, her son, or even her sister. But when she later phoned her husband, she cried as she expressed her feelings to him about how upset she was for their little boy. Then she dried her tears, grabbed a pen and paper, and wrote down the many questions she wanted to ask the doctor the following day, as she started her fight against ML. She had already believed that Sergio's condition was serious, so she wasn't expecting a cure, but she had prayed to God asking Him to help her find a doctor who would give her the answer. Now, she was relieved to be able to move forward in learning all she could about the disease. Although they had planned to stay at the Ronald McDonald House for only five days, expecting that a doctor would give them a diagnosis and send them on their way with instructions for treatments, the visit lasted for almost three months, as treatment was begun.

The doctors knew of no doctors in Venezuela who would be able to treat Sergio's condition, so they recommended that the family move to the United States for the boy to have the best possible quality of life. Everything changed for the entire family that summer, as their mission to care for the youngest child included Sergio's father receiving a job transfer. In 2004, the entire family left everything that was familiar to them and moved to Texas where they started a new life, learning to live one day at a time and loving each other as they deal with the many challenges presented by such a devastating rare disease.

Zachie Haggett

Brenda and John Haggett, of Syracuse NY, felt robbed of the genuine happiness they expected to have when the doctors told them that something was seriously wrong with Zach, their first born, whom they had waited nine years to welcome into the world in 2000. Instead, they were enveloped with fear as to what might be wrong with their seemingly perfect and beautiful little boy as they began a several years long journey to find an answer.

For a few years Zachie made incredible strides with daily visits from Early Intervention therapists, who also provided the support the family needed. It was believed that Zach was only "globally delayed" for unspecific reasons. Brenda and John visited the Genetics Clinic together every six months, where they were told their son looked pretty good, so they could continue on their way and return in six months. There was no reason to expect anything different when the day arrived that John's work schedule precluded his attendance at the visit in May of 2003. It was then that Brenda was told that the doctors saw significant changes in Zach, indicating he might have MPS, and more tests were needed to confirm their suspicions.

The shocking news terrified Brenda, who together with her equally-terrified husband, started researching MPS on the internet and learning that there were many complications related to the incurable disease. The days and weeks of waiting for the results seemed endless, until they finally received the news that the disease was not MPS after all, but it was ML (as with Sergio, the diagnosis was later determined to be ML II/III). They felt somewhat reassured when they were told the condition was not considered to be as severe as many of the MPS conditions.

However, they had the same kind of feelings as most parents, who become, at first, frightened, sad and depressed. As they set out to learn about the new diagnosis, they quickly assumed the position of fighters. Brenda was able to reach another ML mother, who gave them reassurance that they were not totally alone. But, they also were disappointed to learn that there was not much known about the condition, since there are so few families affected by it. Brenda wasted no time in participating with others through the MPS Society and ISMRD, while vowing to do everything she could to make life as happy and successful as possible for her own son, as well as all of the other children she has learned about through these organizations.

In addition to the families profiled here, we have come to know, to love and to admire many more. We try to keep track of these families and to share as much strength and hope as possible with them, for when one member of the family has a rare and debilitating disease, the entire family is touched by that disease. No one escapes the many ramifications and the unrelenting stress produced by such a condition. Everything from nervous breakdowns to marriage break-ups to sibling rivalry and 'acting out' can, and often does, occur. Financial problems can loom with very dark clouds. Parents must balance holding down jobs and caring for their children, as well as finding the help for the children, which is often located many miles away from their homes… a daunting task! The fortunate ones have emotional help in the form of their extended families, their church affiliations and private counseling. It is no wonder that specialized groups such as the MPS Society and ISMRD have grown through the years. They are extremely valuable tools. It has been my privilege, and a fringe benefit of dealing with ML III, to share with the parents I have met through these groups. With admiration, I have witnessed parents and siblings find within themselves courage and compassion they didn't know they possessed, as many of their own future plans have been abandoned, and they have embarked upon a new mission.

CHAPTER THREE

IMMERSED IN THE MEDICAL COMMUNTITY AND ENCOUNTERS WITH PROVIDERS

Most people have some familiarity with doctors, clinics and hospitals. However, parents often find themselves plunged into an unfamiliar and frightening world when suddenly confronted with the diagnosis of their child's rare disease.

During their years of medical training, doctors, nurses, therapists and technicians learn medical terms that become part of their everyday vocabulary. They have no problems communicating with each other when using these terms. Lay people don't always easily understand terms and abbreviations that are common to the medical filed, and parents can feel like they are being left out of conversations when they hear hospital personnel conversing with each other. They often spend the initial months and/or years on a learning curve. As an added complication it can be difficult to ask questions when one is not even sure what those questions should be. Unfortunately some of those questions pop into a parent's head while lying awake in bed in the middle of the night feeling frightened and wondering how they will be able to do all that is necessary for their child.

Even though some medical workers are aware of that fact and are quick to explain things to patients in plain language parents frequently do not receive adequate answers to their questions. And they are surprised by how many different specialists will ask them questions about their child as they repeat the same answers over and over again. While scrambling to learn medical-speak and asking questions of total strangers they face frustrations when they are presented with new tests which add to the fear they harbor of what the results of those tests might mean.

Anyone who is abruptly thrown into an environment, with which they are not familiar, can identify with the feeling of being overwhelmed. These parents are faced with the task of making decisions about a course of treatment while trying to cope with the emotional reactions of confronting a complex rare disease affecting multiple organs. Fortunate are those who receive counsel from geneticists cognizant of these facts, so they carefully explain the meaning of the diagnosis in common language.

Even parents who have a medical background have to learn what the diagnosis of a rare disease with a strange sounding name means to them and their child. Very often there are no adequate answers to the many questions that keep popping up in the minds of the families, producing frustration that only adds to the anxiety they feel.

Kelley Crompton

From the time we have our original encounters we have certain expectations often based upon any previous experiences we have had. My own personal expectations were colored by a time in American history where doctors were portrayed on television programs as being all-caring and all-knowing wise men that could do no wrong. Of course I knew that was not a realistic outlook. Nevertheless, I didn't like accepting less than perfection when it came to caring for my daughter. I wrote about some of our problems with health care workers in my previous book *Kelley's Journey: Facing a Rare Disease with Courage.* But, I'm happy to say that through the years we encountered many more concerned and caring health care providers than we did negative ones, once I learned the value of seeking second and third opinions when it looked like things weren't going well.

In September 1969, I wrote a letter to that wonderful man who had answered our plea for help a year before thanking him for the difference his care had made in Kelley's life and the emotional health of our whole family. I explained that Kelley had grown distrustful of doctors in general, and he had renewed her confidence - mine, too.

During one of Kelley's hospitalizations we met a physical therapist who had a way about her that Kelley responded to very well. If she asked Kelley to do something Kelley said she didn't like this woman would say, "Okay, we'll get back to it later."

Her method apparently made Kelley feel like she had more control over the situation, and she cooperated with this therapist more than she ever had with any other. Thus, I learned a new technique from her. I always tried to learn from the many conscientious and caring professionals we encountered through the years, and I feel only gratitude toward those who saw us through some pretty difficult situations.

Autumn Tobey

Pam Tobey spent years learning everything she could about Juvenile Rheumatoid Arthritis when it was thought that her daughter's symptoms were being caused by that condition. The doctors treating Autumn often said that everything that was happening to her was rare, as it was something they usually would see in older patients – certainly not ten and eleven-year-old children. Even though Pam was told that those symptoms were not typical of JRA she never expected to be informed that Autumn had a disease that she had never heard of. Nothing she had experienced previously had prepared her for the myriad of tests and surgeries that were to come. She credits their particular medical community in Arkansas with being very supportive as their family started on a new path. The hospital staff encouraged parental participation in all of their child's care, including observing some procedures. However, Autumn's dad, David, was not quite prepared to witness the removal of the tube that had been inserted in Autumn's neck for her spinal surgery. When he almost fainted, the surgeon said that David should walk away, since he didn't have time to work on him, too.

When Autumn was a child her parents were always allowed to stay in the hospital rooms with her and beds and recliners were provided. But the first time she had surgery as an adult, they were in the oldest part of the hospital and Pam was devastated that they wouldn't let her stay. Years later the hospital expanded and there was ample room provided for families as well as patients. Pam has also found that most of the medical workers she has encountered have been accepting of the fact that she has learned a great deal about ML III, and they have been open to learning from her. Unfortunately, this has not been the case for all families.

Allison Dennis

Trish Dennis in Australia had no medical knowledge at all prior to Alli's diagnosis. She had been raised believing that doctors were always right and she had no right to question them. She recalls that her perception changed very quickly with Alli's initial diagnosis. As she sat in the doctor's office, he tried to explain that ML III is caused by what he called "a leaking enzyme" so she asked him when they could inject the enzyme to correct the problem, only to learn that such a solution was not possible. Since this was in the days before the internet, there was very little information available. Trish spent hours in hospital libraries trying to find information. She feels that they were fortunate that the hospital provided them visits with a grief counselor to help them get their heads around things after the diagnosis was confirmed.

On the other hand, when Alli was newly diagnosed, they were sent to a major hospital in Sydney to see a rather famous professor. He started speaking about the disorder and was pointing to Alli's little face and comparing her features to those of a gargoyle! The thought of him comparing her beautiful daughter to a cold hard concrete statue horrified Trish. She left his office very upset and refused to ever go back again.

She had read about a professor at a different hospital who seemed to have extensive knowledge on MPS and related disorders so Trish asked the doctors at that hospital if Alli could be referred to him. She was flatly told no! In their eyes it was important that she remain at the hospital with which she was already established. This didn't sit well with her so she searched to find a way to make a transfer happen. Thankfully, one of their previous doctors was willing to write a referral to the specialist they wanted and make the appointment for them. The original hospital staff was not happy, but Trish believed that her daughter deserved the best treatment possible and nothing was going to stop her from providing for Alli's well-being. It turned out to be a very wise thing to do. Alli had many years of wonderful treatment from this doctor and his team.

Another doctor whom Trish regards as an important aspect to her daughter's treatment was her pediatrician. This wonderful man had seen a similar condition in South Africa in his younger days. Therefore, he was very conscious that, due to the rarity of the condition, Alli could very easily be seen as a curiosity. Thus, he protected her from over-servicing. Trish was very grateful for his sensitivity since in the early days of diagnoses the last thing the child needs is to have numerous curious doctors and therapists poking and prodding.

As time went on, Trish learned to carry a notebook in her handbag and write down any questions she had so when they finally did get to see the doctors she wouldn't forget anything. She also asked professionals to explain things in a clear concise manner until she had a firm understanding.

Hayden and Sarah Noble

Paul and Jenny Noble in New Zealand have traveled the world searching for answers for the best ways to care for their children, Hayden and Sarah. During their early years the children were under the care of Pediatric Physicians in Nelson. However, when Hayden was twenty-one and Sarah eighteen years old, the family moved to Tauranga to be closer to other family members who were willing to help out with the very stressful task of caring for two people affected with ML.

Before they moved, their pediatrician from Nelson spent a considerable amount of time talking with the hospital staff in Tauranga explaining the health issues Sarah and Hayden were dealing with and the way to best handle their long term needs. The family had high hopes for a smooth transition with the pediatric department helping them to put in place all of the Adult Services doctors necessary to manage their care. Yet, it took them almost two months to work through the transition to Adult Services. They saw, or more accurately stated – interviewed, several doctors who thought they could manage their care, but they kept making the mistake of comparing Mucolipidosis to other diseases or even calling it by the wrong name. It was disappointing to realize that the doctors had not done their homework and knew nothing about ML. The transition services became a nightmare for the family. It was annoying to discover that doctors in Adult Services believed it wasn't necessary to do follow-up testing, since they felt the previous tests were normal.

The Nobles knew there is nothing normal about the course that ML takes and constantly explained that ML is unpredictable; things can go wrong quickly - often without warning. Some of the doctors either acted as though they knew it all or simply weren't interested and made a diagnosis based on a similar disease. The staff displayed a lack of consideration and little respect for the expertise of the parents, leaving them in total frustration when departing from appointments without resolutions.

In 1990, the Nobles met the doctor who had discovered ML II in 1964. He advised that if they did nothing else, they must manage the cardiac issues that Hayden and Sarah might experience. For the many years that they saw the Pediatric physicians, annual testing had been conducted. It was during one of these annual tests that a hole was found in Sarah's heart and they were advised to keep an eye on it.

By 2009, the Nobles had spent six years advocating with Adult Services in Tauranga. Even though Sarah had scans done she still had not met with a cardiologist when she started experiencing shortness of breath, chest pains and chronic tiredness. They knew these symptoms could be related to the hole in Sarah's heart that had been previously diagnosed, but they had no one to talk to about those symptoms. Her Echo tests were never explained to them and the report they received only stated that there was no change from her last scan, leaving the family to wonder what that meant.

Sarah became very concerned about her chest pains when they became more regular, so Jenny decided it was time to insist that a cardiologist see her. Since it's often difficult and frustrating to advocate for a family member, Jenny made an appointment with their General Practitioner. She outlined Sarah's symptoms, the issues of access to the cardiac team, and the lack of real results from Sarah's Echo tests. Acknowledging that her treatment at that hospital was unacceptable, the doctor made an urgent request to have Sarah seen by a cardiologist.

Within two weeks Jenny and Sarah were sitting in front of a cardiologist at the Tauranga hospital. That consultation became extremely difficult when the doctor turned his back on Jenny and spoke only to Sarah. As he was writing notes, Jenny whispered to Sarah that she had forgotten to tell him about her chest pains. He quickly turned to Jenny and said he would not be discussing the matter with her since she was not the patient.

Mother and daughter were both pretty shocked, but Sarah rose to the occasion! While examining her, the doctor asked a question about ML, so she told him he needed to speak to her mother. Still, he refused to do that, telling Jenny he would only be guided by medical papers and requesting that she supply them.

After a contentious discussion, Jenny bought up the possibility that the hole in Sarah's heart that had been discovered years before was still there and causing her symptoms. The doctor said there was nothing in her notes about a hole and that they had never found it on scans. Jenny said that if he looked at the scans from the previous hospital he would see it. The doctor was not convinced. However, he did decide that Sarah should wear a cardiac monitor for two weeks. That was organized quite quickly, and they received an appointment for a Cardiac MRI to be done in Auckland.

That November, after Sarah had the MRI, Jenny was sitting in the waiting room where she could hear the doctor talking to the doctor in Tauranga. He stated that he had found a hole at the top of Sarah's heart and that the vein from her lungs was pumping into both right and left sides of her heart, causing the right side of her heart to be enlarged.

What the Nobles didn't know was that the hole in her heart was there from birth and these holes often don't affect people until they are much older. However, because of the ML, the hole affected Sarah sooner than it would have in a normal person.

Sarah finally had open-heart surgery in August, 2010 at the Auckland Hospital, ending up in Pediatric ICU. The Nobles were relieved and pleased that she received wonderful care by the Pediatric Team involved at that time. They made sure that Paul and Jenny were totally involved, and assured them that, going forward, should Sarah have any other problems with her heart, such as the valves getting more damaged and needing repair or replacement, she would have it done through Pediatric Services. Sarah was excited when they gave her a card with all the contact details to carry with her at all times. The family felt this to be a major breakthrough, knowing she would now get the right care for her heart.

Sarah had a six-week post-operation appointment in September 2010 at the Tauranga Hospital. The doctor who had given them all the grief in the beginning had a complete change of attitude. He not only spoke with Sarah, but also included Jenny in the conversations. While she doesn't expect to ever get an apology, Jenny was relieved that he finally showed some respect and understanding of the knowledge the family has of ML.

When the doctor discussed management of Sarah's heart going forward, Jenny was able to say that they needed to add Hayden into the mix as well. To their happy surprise, he agreed to work out a long-term care plan.

Sarah's chronic tiredness, shortness of breath, and palpitations due to the heart defect disappeared once she recovered from the surgery. She received a new quality of life, and the doctor seemed to be very pleased and surprised that results were being seen so quickly. Jenny always believed that would be the case, as the international experts they met told them that once the hole was closed Sarah would notice the changes quite quickly.

Jenny contends that living with a rare disease is probably one of life's hardest journeys for parents who have affected children. She feels strongly that there needs to be more respect of this knowledge by the professionals caring for these children, since the parents are almost always the experts.

Huddy and Sammy Anthony

Liz and Tom Anthony had no 'heavy-duty' involvement with the medical community until the time they needed help for their boys. From the outset, most of their experiences with specialists, pediatricians, nurses, therapists, medical supply providers, and lab technicians were positive. Liz speaks of being thankful for all of the healthcare workers she has encountered. Most of the doctors they have met have been friendly and have treated the family with quality care.

As a mom to two special needs children, Liz had to deal almost daily with medical people starting before the diagnosis was made. Her boys had over ten years of private Physical Therapy and Occupational Therapy, plus several years of Speech and Hydrotherapy. At one point, for at least a year, they had therapy four times weekly, and the therapists became like family members. Hudson's first physical therapist was a no-nonsense, down to brass tacks woman who worked Hudson over for an entire hour each session while he cried nonstop and Liz cringed at the 'torture' sessions. However, if it had not been for the aggressive approach, Hudson might not have learned to crawl. One exceptional Occupational Therapist was especially personable and got to know Huddy and Sammy's likes and dislikes. Another therapist was extremely bubbly, and made a game out of learning how to speak and pronounce sounds. There were many more excellent therapists who knew every muscle and how to get them going, and they accomplished the job in a fun, creative way. All in all, Liz credits the therapists helping Sammy and Huddy throughout the years as being absolutely outstanding in knowing their stuff. She found that the therapists who are also moms can encourage any child in therapy unusually well.

Dental visits for the boys were not normal since they could not sit normally nor give clear x-rays. When two teeth needed to be pulled for Sammy, because there were more teeth under the gums, they were able to see a wonderful dentist at a Children's Hospital Special Dental Clinic. Sammy's oxygen level and heartbeat were monitored while his special-needs dentist, who resembled a linebacker, capably removed his two molars under local anesthesia, shouting "Yeehaw!"

The recovery from each of those extractions took about a month. Sammy had a slow recovery from the pain, and he needed a children's analgesic for many days. Huddy has a resorbing tooth which has remained in his mouth for a few years since the dentist found it. They each had sealants put on their teeth and have had only one or two cavities.

Sammy and Huddy both had pneumonia and the flu at about the same time, but only Sammy had a four-day hospitalization because of his compromised pulmonary system. In the hospital he let everyone know how much he disliked "pokes", IVs, oxygen masks and the BiPAP mask. He would ask, "What do I get for putting up with this?"

Liz was grateful that she was able to stay with him around the clock in the hospital, because he wanted her there, and it was doable. Huddy stayed at his grandma's during that time so Tom could go to work. They found that having a child in the hospital caused a major disruption in their daily routine. That was Sammy's only hospitalization except for his several sleep studies.

Callie Nagle

Callie's parents, Richard and Debbie, were introduced to the medical community quite a while before they learned that she had a diagnosis of ML III. Their first experience came about when Callie was only eighteen-months-old. Debbie was giving her baby a bath when she noticed a bulge right below her little belly, causing her to "freak out". She immediately pulled the baby out of the tub, dressed her and called their pediatrician. He saw them right away. However, when he examined Callie the bulge had disappeared so Debbie started feeling a little crazy and uncertain. The pediatrician knew the bulge had been caused by an inguinal hernia, so he referred them to a surgeon. The local surgeon they saw next had such a horrible and cold bedside manner that Debbie remembers saying she wouldn't bring her dog to be treated by him. In recalling that incident, Debbie says she thinks it was a silly remark because their dog is also a part of their family. Callie was screaming during the examination and he never so much as looked at her or tried to soothe her, but only seemed annoyed that she was crying. He suggested surgically repairing both sides even though she was only presenting symptoms on one side.

Like many other families, they sought out another surgeon and met with success. They felt confident with him from the first meeting. He was kind, sincere and caring. Although Callie was scheduled for outpatient surgery it was a scary time for her parents knowing that their little girl was going under anesthesia. The first time they handed her over to the hospital staff, when it was time for her surgery, proved to be a memorable one. They gave Callie a medicated lollipop and a small drink prior to going in to sedate her, but met with no luck. They were all amazed that these 'tricks' had no effect on Callie. Debbie finds it impossible to describe the sick feeling she had when Callie went off in a stranger's arms while crying for her mommy. That hospital's policy didn't allow parents in the operating room. They had to go through this one more time, when, at age three, she presented with another inguinal hernia. It was amazing to Richard and Debbie that, when they returned home after those operations, Callie ran around as if nothing had happened. They were not only amazed, but also relieved.

Jennifer Klein

Terri Klein had extensive previous experience with the medical community long before her daughter Jennifer was born. When Terri was eighteen years old, her thirty-year-old sister-in-law, Nancy, was diagnosed with breast cancer. Terri became intensely involved for over a year in helping to take care of Nancy as well as Nancy's children. She learned what it is like to hold the hand of a cancer patient who was suffering from radiation and chemotherapy. It was during that time that she learned medical terminology as well as valuable lessons about patient advocacy as she watched the way in which her mother argued with, and sometimes yelled at, doctors when Nancy was unable to speak for herself. This was the time she also learned that, although doctors are trying to help, they don't always have the necessary answers. Unfortunately, Nancy died despite their efforts. Shortly after that, Terri's mother slipped and fell on a marble floor and spent two years in a wheelchair. She recovered from that, and then Terri's father had a massive heart attack. Then Terri's mother, who had asthma and was hospitalized often, was diagnosed with Amyotrophic Lateral Sclerosis (Lou Gehrig's disease), subsequently dying at the age of Sixty, just before the family received the correct diagnosis for Jenny's physical problems.

From her previous experiences, Terri learned that, "you don't know what you don't know, and you don't even know what you are supposed to be asking." Jenny's original diagnosis was incorrect, because the diagnosis had been made based upon what was seen in her X-rays. Further testing led to the discovery that Jenny actually had ML III. It was only after learning about this progressive Lysosomal Storage Disease, and the way in which bodies are affected by it, that they understood the remark by an anesthesiologist when Jenny had surgery, "Boy, she had a tricky airway."

Andre Andrews

When Andre had pneumonia before he was a year old, his mother, Jane, had just started a new job but just the thought of him being in the hospital at that tender age was unnerving to her! Fortunately the hospital was right across the street from her job so she virtually lived between the hospital and the job. Once he was admitted, she stayed with him nightly until he fell asleep. She tried to be at the hospital when he awoke each morning to feed him breakfast and stay with him until he was comfortably settled, although she felt that children really aren't easy to settle at that age. He fretted when she left, but she returned for lunch, and then again for dinner, remaining until he fell asleep. That was her protocol until he was discharged a week later. She had no idea at that time that they were to have many more hospital encounters, since Andre had not yet been diagnosed with ML II.

Jane came from a medical background. She did transcriptions in a hospital covering a variety of medical specialties, but she never came across terms like Lysosomal Storage, Mucolipidosis, or I-Cell (ML II). Before Andre's diagnosis, when she heard about genetics and gene testing, she thought only of DNA and paternity and not how a missing cell from one parent with a similar cell from the other parent can cause such a rare disorder.

She doesn't think that the information she was given at the time of diagnosis was adequate. She is amazed as to how much she has had to learn as a continuous process. Like many other parents, Jane has had to educate the providers after receiving the information from the MPS Society. She thinks that she still doesn't speak the language as well as she should, or if she is able to explain it was well as some other parents, but she does the best she can. Jane believes that having had the opportunity to raise Andre has given her a totally different perspective of the medical world, and she thinks it's sad that not everyone is given this opportunity. It's not just the medical world; it's the world as a whole.

On an annual basis, Jane has been able to meet with Andre's geneticist, orthopedist and physical medicine team to discuss his condition. Through the years, Andre's examinations and tests have been many, such as an MRI of the brain with and without contrast as well as the cervical and lumbar spine. Since this actually totals eight procedures, and they can only do two at a time, he has them spread out over a month. He has had CT scans, sleep studies, dexa scans, and testing for carpal tunnel syndrome. They saw a nutritionist to check body fat and Andre fell within the 50 percentile. Andre has seen an Oral Surgeon, since extra teeth can cause problems. And he has had problems with his ears, requiring tubes to assist in drainage. There have been times that Jane has vacillated between thinking that sometimes children like Andre see enough specialists, yet, perhaps they can't see too many. She always wants to be reassured that they have covered all the bases.

When Andre was eleven years old, he went for his first Aquatic Therapy session. Jane felt as if his reaction to that experience would give her (and everyone else in the room) a heart attack. She never expected him to react as badly as he did. She thought she had prepared him by telling him what she thought the experience would be like. The only time previously he really got in the pool was when they went to "Give Kids the World" and he sat and dangled his feet along the sides. But this was like an oversized fish tank that one could see clear through. It could have been that was what frightened him. Jane insisted that he at least try it but it took some prodding.

Andre finally decided to let the therapist hold him and walk around the pool with him in her arms. She managed to get him to make circular motions with his arms and then to kick his feet. Jane had to stand along-side the pool to calm him down. They had a little flotation device for him to sit in. He screamed to the top of his lungs but they managed to get him in it anyway. He wanted her to hold the raft, and she did while he was holding onto her ever so tightly at the end of the tank. She let go of it and held his hands until finally she was able to get him to do some moving. Andre eventually settled down and actually kicked his little legs and became slightly more comfortable as the session went on. They showed him how he could play basketball but he still wasn't thrilled. Yet, by the end of the session he was actually looking forward to the following week. They took him out and gave him a shower that helped him to relax while the hose sprayed warm water all over him. Andre was set up for hourly sessions for the following month. When the therapist with whom he was comfortable left, it was a challenge to try to encourage him to continue, so Jane didn't push the issue at that time. Years later he returned to aquatic therapy and seems to be enjoying it. He has befriended a new therapist and is content.

Joey Nagy

Because Joey Nagy's symptoms showed up so early in his life, his parents, Linda and Frank, were involved with the health care community from the time he was born. Yet it took them years to find the correct diagnosis and move on to visits with one doctor after another to ascertain baselines through testing. The results of the testing indicated that Joey might need tubes in the ears since the fluid that would not drain could affect his hearing. There was also a suggestion that he might need to have his tonsils and adenoids removed, and they also were advised to test for Carpal Tunnel Syndrome.

At times they ran into doctors who didn't understand ML III at all. One talked about I-Cell, which is ML II, not ML III. Luckily, shortly after they received the diagnosis, the family was able to attend a conference where they were relieved and delighted to meet a few doctors who are experts in the field of Mucolipidosis. The information they received about the disease proved to be invaluable. The doctors advised them that Joey should not have hip surgery at that time and should not engage in some types of physical therapy. They learned it would be best for them to take Joey out of regular gym classes at school. Linda thinks that she might have gone ahead with less beneficial recommendations from others if it hadn't been for this help. After they returned home she was able to tap into a wealth of knowledge from one of the doctors via email.

They also made a beneficial connection with a genetic doctor who took a real interest in Joey, even traveling to attend a conference where she could also meet up with the experts. Linda had never imagined that a doctor would always be there for her, promptly responding to her needs. The family made a decision to have all of Joey's medical care handled at the same hospital at that time, since his condition is so complex and it would more easily facilitate correspondence between the doctors.

In January of 2008, Joey was almost thirteen years old when it finally became necessary for him to have surgery for his hips. The following account is descriptive of the many types of situations that patients run into when dealing with numerous medical systems.

When Joey had recovered sufficiently from the surgery to be discharged from the hospital, he needed months of rehabilitation before he was able to return to his home. The way in which the progress took place had them feeling like they were riding on a roller coaster. Linda likened her role at that time to that of a Ring Master at a circus, trying to direct various rings. First they were told that Joey was going directly to rehab. Then they were told he was not. Yet, a short time later, they were told he was. When it looked as if it was a 'go' again, they were suddenly told that there was a set of four different criteria that Joey had to meet in order to be covered by insurance. According to the Physical Therapist he met all the criteria. Joey needed to be able to do upper extremity therapy, isometric exercises, be able to go in a reclining wheelchair and finally, be able to be placed on a tilt board. They were ready to write the order when someone from the orthopedic department said they didn't like the tilt board. Also, Joey had spiked a fever the previous night, and they were told that he had to be fever-free for twenty-four hours.

The next time they were ready to go to rehab, the fourth criteria had been changed. Now, Joey needed not only to be able to go on a transition board, but to also to get from the bed to the wheelchair. The following day they actually made some progress. Linda was able to help the two physical therapists partially sit him up at the side of the bed, but it took them almost forty-five minutes to get him into that position. The following day they worked to assist him into the reclining wheelchair, so the rehab would accept him. During this time Linda began to feel like a robot, just going through the motions, feeling unconnected in the hospital environment. She looked forward to Joey moving on, but she was also afraid to get her hopes up because of the way things kept changing.

Finally, their hospital caseworker did a great deal of work to make the transition a reality. And when Linda arrived at the hospital, she was pleased to see that the physical therapists had Joey sitting up again. They moved him to the reclining wheelchair, allowing Linda and Joey to take off for a tour of the hospital. Linda wheeled him down to the main floor where they leisurely ate their breakfast in the lobby, and then were off to see the Family Life Center. They found it to be an amazing place where families and kids can relax and have fun.

Once they were ready to release Joey from the hospital, Linda couldn't believe it was really happening. The rehab felt like a Five Star Hotel compared to the hospital. The staff got him washed up and dressed in his Tony the Tiger lounge pants.

Joey hadn't been doing any exercises for a long time, so the therapists set out to change that, and work him hard. Like most parents, Linda thought that it was better them than her. She didn't like to see him suffer.

With the help of the wonderful staff, Joey made great progress. They keep him busy with physical, occupational and recreational therapies, alternating all day long. Every day, he looked and felt a little better. He also managed to charm the nurses. At the end of the day, or when they did something nice for him, he asked them for hugs. They all seemed to adore him.

They had to wait for a couple of wounds to heal before he could go in their warm water therapy pool. There were goals that Joey needed to achieve before moving on to the next level. If they were not met, and if he couldn't progress to the next level of goals, then they wouldn't keep him there and he would be sent home. An ambulance was required when it was time for Joey to visit his orthopedic doctor, who determined that Joey could progress to weight bearing on his legs.

Staffing meetings were held every Tuesday to discuss patient progress and possible release. As much as Linda wanted to have him home, she felt the rehab was best possible place for Joey. He did progress, and they settled into a routine until the time that Joey was ready to return home.

Linda discovered that she had never realized the situation would be so hard on the entire family. Frank missed them, and visited the rehab in the evenings. Even Frankie, who was sixteen-years-old at that time, said he missed them. But Linda suspected that her teenage son really missed the meals he was used to getting.

Anna James

When Anna was first diagnosed, her parents didn't know anything! They had basically taken her for her regular pediatric checkups.

Once Bret and Jackie attended that ominous meeting with the diagnostic specialist, their world changed forever. The overwhelming visit included a complete day of check-ups with various doctors - neurologists, radiology, blood work, urine, orthopedic and more. Anna handled it well, despite being quite worn out. During those initial days of discovery, it honestly sounded to them like the healthcare workers were speaking a different language. Much of the information they received at that time did not sink in until later. It was a truly terrible feeling. Jackie is horrified when looking back to remember that Anna was misdiagnosed because they didn't wait for the results of blood-work and other information that would be used to correctly diagnose her. They lived through six weeks of utter hell before receiving the correct diagnosis of Mucolipidosis III. They had been told that their child was going to die in a terrible way, and then they were just left to deal with it by themselves. They were not even told the news in a way that made it relatively understandable. The medical jargon that was used made it very difficult for them to comprehend what they were being told. Some of the doctors they encountered during this time were very clinical and matter of fact. Jackie and Bret didn't really know what they were testing for.

However, there was one doctor that gave them some hope. When they took Anna to visit an oncologist at a Children's Hospital, they were told that one of the possible treatments to hopefully lengthen Anna's lifespan at the time was a bone marrow transplant. This doctor explained very carefully what a bone marrow transplant would mean, making everything very understandable. He then went on to say that he had treated MPS patients before in the past, and in his opinion he did not believe Anna had MPS. That gave them a tiny prick of hope.

It was a few days before Christmas, when they got the best Christmas gift possible, and they had their first appointment with the geneticist. Jackie still can't believe they had not been sent to visit a geneticist first! That wonderful doctor began talking about a disease called Mucolipidosis. After a minute or so Jackie stopped him and asked him what he was talking about, since they had been told Anna had MPS. The geneticist was quite taken aback. He was surprised that they had been given a diagnosis, since it was his job to do that - once all the blood-work and tests had come back. He did say that both disorders were very similar and that he could see why they may have thought she had MPS. But he didn't say what Jackie thought, which was, "Why would they give a definite diagnosis of something, when they didn't know for sure?"

Their geneticist was a wonderful man, explaining everything in detail in layman's terms. If they didn't understand anything, he was more than happy to explain. He also had a wonderful assistant that they could call for answers to their many questions. The family was sad when that geneticist left to work at another hospital after treating Anna for eight years.

There were a few doctors that they encountered that were very difficult to understand or sometimes even appeared to be very brusque. They actually have changed doctors a couple of times through the years, and Jackie found that she had no problem in telling the receptionist that she wanted a change of doctor. Yet, even though there have been a couple of instances where she felt that her initial visit with specific doctors did not go too well, and she was uncomfortable with them, some of those doctors turned out to be her favorite doctors, once she moved out of her comfort zone and confronted situations! Jackie learned that it is important for her to be bold and speak out, even though she hated to do so. She would much rather be compliant and easy to deal with. Today she would advise any parent with a special needs child to never ever become compliant and never ever be afraid to speak up if they don't understand what a doctor is saying. Furthermore, they should not move forward with any procedure if they are not comfortable about it. All parents need to ask questions and push until they are comfortable with the answers to their questions. Jackie has learned that it can be done in a polite, but very firm, manner. Parents need to let the physician know that they are their child's biggest advocate. Jackie speaks for many other parents when she says that she does get tired of trying to push for everything. We all often wonder why it is that we have to push for so many of the services our children need, leading us to complete frustration! As a mother, Jackie feels that the biggest hurt is that we can't just wipe away the tears, kiss away their pain, and have them get up and be all better.

Spencer Gates

From the time that the Gates family started dealing with the medical community, they ran into a variety of unexpected circumstances. Yet, their experiences were not unlike those of other children with rare diseases, when many tests need to be performed. Andrea feels that it is almost a miracle that they all survive without major complications.

Spencer appeared to be doing fine during his bone density test where bags filled with rice were laid on his legs that had been strapped into a frame. But, as they were leaving the office Spencer said his legs hurt so much that they were at a twenty! He was referring to the usual pain scale of one-to-ten. He said that he knew the scale didn't go up as high as twenty, but that was how bad his legs hurt. Andrea felt terrible because she had no idea he had been in pain! He told her that the heavy bags and the straps made his legs bend straight for a long time, which is a very unnatural position for a child who really can't straighten his legs. It had never occurred to Andrea that the weight of the bags would be so heavy. His mother gave him credit for going through with the entire test, and told him that he was a real true trooper. Spencer further asked if it was alright to add the number twenty to the scale when his legs hurt so badly. Andrea was almost in tears when she said, "Of course!" They ended up compromising on keeping the one-to-ten scale with fifteen and twenty used as necessary.

At an eye appointment with a doctor who had been seeing Spencer for about six years, the doctor performed a very quick eye exam and stated that Spencer would be fine for a few more weeks. Then he told them to return in a few weeks to have his eyes dilated and get a new prescription. Andrea was a little dazed as they left. When they were a block away she pondered why they would have to go back two weeks later. So went back to ask if everything could be done that day, in the way it always had been done in the past. She learned that the insurance company had not approved that exam. She asked them to call someone at the insurance office to get the correct authorization, and she also called Spencer's doctor to help speed up the process. Four hours later (with no food for the kids), Andrea was frustrated. Spencer's doctor's secretary told them that she knew it would get approved, as she just needed the correct code, so they could go ahead and do the exam and she would call with the authorization as soon as she got it. The person who handled the authorizations said she had to go to lunch and would work on it when she got back. By that time, Andrea's head was spinning and the kids were hungry. They left. Andrea called an hour later to see if the authorization had come in. It had not. A few minutes later, Spencer's doctor's office called to say she had just faxed the authorization to the eye doctor. So Andrea called to see if they received it and was told they did. She said that they could be there in five minutes. But then the doctor was running behind, so he could not see Spencer that day. They offered to set up a new appointment, which, of course, would require new copay. Andrea found it hard to believe, although in another way, it seemed almost comical that she had the kids in that office for four hours. That's what really bothered her as she thought that doctors sometimes forget that they are dealing with a person and not just a disease.

On another occasion, they received a message from Spencer's geneticist to reschedule the appointment that took two months to obtain. It didn't surprise her that she had problems getting through to reschedule. Once again, she felt frustrated that they were being put off for months.

When the doctors recommended hydrotherapy for Spencer, the insurance company denied payment by simply sending a letter stating that an appeal could be filed. Working with one of Spencer's therapists, a call was placed to an orthopedic doctor through a local foundation that provides hydrotherapy. Andrea was relieved when he said Spencer would be a great candidate for their services. Nothing was said about payments when Spencer was sent for an evaluation. The Physical Therapist said she could start seeing him twice a week, starting immediately. Andrea was thrilled, but unsure about payment until she was handed a paper that stated they charge $20 a month to people that could afford to pay. That was a major relief! They went back later that day and Spencer enjoyed the pool with a very nice a therapist who was great with him.

They also ran into problems arranging a test to determine if Spencer had Carpel Tunnel Syndrome. It took two weeks to get in touch with the woman who made appointments. When she did get back to Andrea, she told her that she needed more information before she could even make an appointment. Andrea felt like they were trying to drive her crazy with the endless delays in getting Spencer's tests done.

The Pulmonologist who saw Spencer ordered a series of tests. Andrea asked the doctor and nurse what most of those tests were, but they both talked too fast for her to understand after being there for almost three hours. They all wanted to go home. In addition to tests for his lungs and heart, a sleep study was ordered.

At the dentist, Spencer's x-ray revealed that his molars were growing sideways. Two other teeth were growing very premature with no space underneath them, and two teeth were growing at a weird angle, causing the two teeth next to them to also grow at a weird angle. The dentist said that one good hit could break the jaw! It was a bit unnerving to Andrea dentist kept saying that this was an interesting case.

Andrea's experience has been that it is hard to know what will transpire when dealing with doctors. Sometimes, when she has brought up something that was discussed at the last visit, the doctor has looked at her with a blank face as if she was saying it for the first time ever. She has thought about taking a tape recorder along with all the reports and tests that had been ordered from the last visit.

On the positive side, they really liked the neurologist Spencer saw because he never had a problem acknowledging that Andrea and Kevin knew more about ML than he did. They found that to be very refreshing.

Sergio Cardenas

The situation was very different for Maria Elena Cardenas, mother of Sergio. In the days before they had a diagnosis, when they were still in Venezuela, she not only needed to learn about the medical terminology, but she also was forced into a crash course on nursing. There was a national labor strike throughout their entire country, which was very scary, with no hospitals, pharmacies, gas stations or grocery stores open. There were almost no services of any kind, and they lived in a very small town where pediatric specialists didn't exist. So Maria Elena learned how to use needles to draw blood; set up and administer I.V. medicines, and perform all of the other duties of a nurse. During the time when no hospital services were available, Sergio suffered from pneumonia frequently. Maria Elena felt lucky that he survived.

Later on, when they returned home from their trip to the US to receive a diagnosis, Maria Elena had an occasion to take Sergio to the Emergency Room in Venezuela. She was stunned when they told her that they didn't know how to care for someone with such a rare disease, so she should go back to the US to get help. Frustrating days followed! If she had questions, she had to write or call a doctor in the US to ask them and wait for the answers. She didn't know any English at that time, so she had to ask the questions in Spanish and hope that she would receive an answer. When the answer did arrive, it was in English. There were times when questions were left unanswered. Sergio's pediatrician tried to be helpful. She asked Maria Elena for the booklet she had received in the US so she could learn more about ML. In searching for answers on the internet, they realized they were spending many hours at the computer researching about ML and special children, but everything was in English and difficult to translate.

Maria Elena and Gustavo became very discouraged when they looked for treatment for Sergio's symptoms in their country. After praying for guidance, they came to the conclusion that the only way they could receive the best care available for the boy would be to move the family to the US. Their decision was clear. Thankfully, Sergio's father was able to obtain a transfer with the company for which he worked, and the family moved to Texas when Sergio was four years old. Although they have experienced some unpleasant encounters with the health care system in the US, Maria Elena prefers to focus on the beneficial treatments Sergio has received, and the fact that he started having a better life once they had relocated to the US.

Zachie Haggett

From the time they learned their son had ML, Brenda and John searched for information, attending conferences and meeting with other families dealing with the same condition. They rode an emotional rollercoaster while coming to grips with what the diagnosis meant for their future. They approached their first conference hoping to find information about bone marrow transplantation, only to learn that it was not recommended for Zachie. Although disappointed, they met several families living with ML, and doing quite well at that. They had hope again, becoming fast friends and confidantes with the families who truly understood all of the emotions and fears they were going through and returned home with a mixed bag of disappointment and hope as well as many new phone numbers.

Shortly after his seventh birthday, the doctors wanted to ascertain the severity of Zachie's kyphosis. A CT Scan and an MRI were scheduled, for which anesthesia was required, meaning that Zachie had to be intubated.

Having educated themselves about what could go wrong, Brenda and John were very anxious. They prayed that the anesthesiologist and all who would be in that room would be very cautious and have God's hand and guidance with them so Zach would do well.

Zachie was bubbly that morning, being excited to go to the hospital where he had many friends. There was some confusion when he was taken in, then brought back, and then taken in again with different staff members making opposing decisions as to the order things should be done. This didn't help to calm any of them, and the little boy cried and screamed until he fell asleep. Then the wait began for his parents.

Zachie's airway had already become so obstructed that the anesthesiologist said she was nervous; that was comforting, as the slightest mistake could end his life. Brenda asked if she would be using the fiberoptic tube to intubate, and she said that they would. She also assured them that as soon as she got him intubated she would inform them. So they waited… and waited… and waited, while becoming increasingly nervous. A very long forty-five minutes later, she reported that Zachie's airway was much worse than it had been a year before, and that she also kept getting stuck at some large lumpy thing at the base of his throat. She didn't know what it was, so she asked if it would be okay to have Zach's ENT doctor take a look. They agreed and nervously waited again.

The tests were to take about ninety minutes, but two hours later, they saw Zachie being transported from the CT room to the MRI room along with an entourage of people and a lot of beeping equipment. Feeling more uncomfortable, they proceeded on to another waiting room. Finally, the Anesthesiologist told them that they were going to take him to an OR to have the ENT doctor look at him. Forty five minutes later, they saw the ENT doctor enter the waiting room, and sit down with another couple. Obviously he had not yet seen Zachie.

When he did address the Haggetts, he appeared to be in a hurry. It seemed like he didn't want to waste time talking to them, but they asked what his plan was. He said that the anesthesiologist thought she saw something, so he'd go take a quick look with the scope, but that there was probably "nothing to see."

Feeling that he was taking neither them nor the anesthesiologist seriously, they asked him to please take his time, and go slow. He rolled his eyes as he said he'd let them know, and he left.

After another forty-five minutes both he and the anesthesiologist returned appearing flushed. It was disconcerting that he went from room to room trying to find a private spot. He finally just stopped in the doorway to the waiting room and began his report.

"She was right," he said, "There is something down there." He went on to explain that with all the storage and the airway being completely swollen, he could not get close enough to see what it was, and at that very moment, they were bagging pressure to him because his airway was almost closed. Then he requested their consent to do a tracheotomy on Zachie!

Being shocked, Brenda instantly said "No! You can't trach him! He will not make it!" This is exactly the scenario that had taken place with a number of ML children that didn't survive, due to their fragile airways. They had warned the doctors to proceed cautiously so they would not end up in that situation! The doctor rolled his eyes again telling them that he didn't see how that would happen and that a trach was the only option. John and Brenda both tried to make him understand that it was not an option; he needed to figure out how to keep that airway open with the tubes and not trach him. He threw up his hands and said, "I don't know what you want me to do, he's not going to make it if we don't trach him." They responded that he would not make it if he did trach him! Then, to their complete amazement, he said, "I have had experience with a 'Hurler' child and a tracheotomy."

Shaking and crying, Brenda vehemently declared, "He does not have Hurlers! That's MPS. Zachie has ML! They are not the same! ML kids are even more fragile than Hurler babies and are they are also rarer! You have not seen this before! You cannot do this! If he did somehow survive the trach, he would absolutely die inside from frustration. Anyone who has met him knows that Zachie is funny and smart and he is all that he is because of his little voice. It would kill him to take that away! He's had enough taken away from already. He can't play like other kids or do the things other "normal" kids do. All he has is his Character, that is who he is and we love that! He loves that! We could never strip him of that! It would not be a life for him. And that is only if he actually made it through. We know of a few who have survived, but most are not done in emergency situations; they are planned and done with extreme caution!"

She was thinking that this doctor was trying to cowboy his way through the situation with a child who he didn't even know. He didn't even know the correct diagnosis, even though he had been one of Zachie's doctors for five years!

Knowing that they must advocate for their son, they told the doctor to do everything possible to avoid a trach. As they left, the anesthesiologist gave them one last hug, while they thanked God that she was with them. The next two hours were very long. They were left alone, with no one even updating them during that time. Finally, after three inquiries, they were told that the doctors weren't doing anything until the medicine wore off; they were standing-by because Zachie's sats were very low and the scope was still in. After another agonizing hour, a nurse informed them that he was in recovery, and they could see him. They were incredulous that neither doctor gave them that information. They asked whether or not he had a trach. She said they didn't have to trach him, but they left all the tubes in for the time being. Zachie's parents finally felt like they could breathe again, and let the tears of relief flow.

When the ENT doctor did show up, he said, "Well, that was fun." By that time, Brenda was so sick-to-death with his careless attitude she promptly told him that she hoped he would never again forget what ML looked like!

The anesthesiologist acknowledged that the Haggetts were right, it was worse, and their concerns were valid. She said she was glad they had taken so much time to prepare them. However, they believed that the ENT doctor didn't respect them or their opinion, and he was ultimately the one in charge at the moment of truth. They remained shocked at the way he handled the situation.

Once in the recovery room, they found a very bloated Zachie. His nose and mouth were bloody, since they had unsuccessfully tried to remove one of his front teeth to make more room for the tube. When he awakened a little, he screamed and moaned because he had two tubes down his nose and through his throat. His foot was bandaged up with an IV and he was thrashing all over. He tried to talk, but sounded scratchy from the tube, so they couldn't understand him. It took several hours for him to really awaken. He was under for a total of six hours.

In the ICU Zachie slowly became aware, and the more aware he was, the madder he became that he had tubes in his nose. He screamed and thrashed, but awakened every hour, and asked if it was time to go to McDonalds yet. Brenda slept in a chair and John slept while holding him in the bed. The anesthesiologist checked on him many times throughout the night. Morning arrived, bringing with it rounds of people including all those involved in the previous day's fiasco. They wanted to scope his throat again to make sure that the airway was better before he could be discharged. John and Brenda were not happy with the idea of torturing Zachie once again with the scope, but they had to make sure he was not still so swollen. The minute they rolled Zachie into the office of the ENT, he began to cry, knowing where they were and what was coming. A bronchoscope is a very difficult and uncomfortable procedure. Brenda felt awful. Normally she could be strong when he had this done, but she left the room so she wouldn't cry in front of him.

The ENT doctor told Zachie to "settle down and stop squirming!" Then he got a full view of his throat and airway and saw the polyp at the base. His airway looked very good, still small but not as swollen. The entire event being concluded, the ENT doctor said, "Well everything looks good so he should be fine."

Brenda responded, "So it's a good thing we said no to you yesterday then, huh?"

He said, "Well, you know, it was hairy in there yesterday and we really didn't think we were going to get the swelling down. Honestly, you are going to have to reconsider the trach if he's going to have to do more surgeries.

Additionally, I would not have started with such a large tube to intubate. I would have used a fiberoptic from the get-go."

They couldn't believe what they heard him say! He actually admitted that he did not listen to Zachie's parents when they said that they must use a fiberoptic tube to intubate Zach because his airway is so small and fragile. And then, after it happened and they discussed why they didn't want a trach, he continued to think they should! They wondered what the doctors were doing when they explaining this to them. Were they only pretending to listen?

Brenda finds it difficult to accurately explain how upset she and John were. First, they did not think Zachie was going to make it out of the OR, and they actually discussed losing him. Then they were furious that they were not listened to, even though, as parents of a child with one of the rarest diseases in the world, they had spent four years educating themselves and meeting the doctors who actually deal with these diseases, so that they could bring the most recent information to the doctors who treat him. They found the arrogance of some doctors to be unbelievably disturbing and wondered how a doctor could be so heartless as to leave them in that waiting room for hours not telling them whether or not Zachie was even still alive, or if he had to trach him, nothing! When Brenda later spoke to his pediatrician, she was equally disturbed, and concurred that such incidents really should not happen.

CHAPTER FOUR

FINDING SUPPORT

Once a child's diagnosis is confirmed, parents want to learn all they can about the disease. Years ago little information was available about most rare diseases. Technology is a wonderful tool, not only for researchers, but also for families. It doesn't take very long today to make the right connections to find others who are dealing with the same rare disease. Relief is the first reaction families have when someone responds to their call for help, being reassured that someone else knows what they are talking about. Relationships are quickly formed, as an instant understanding is found when they connect with each other, and help is generously offered. Some parents become very active and form organizations in order to look for funding for research here, there, and everywhere.

With the advent of the internet, some ML families started to connect with each other after listing our email addresses with the MPS Society. However, the message I received in July 2003 from Paul Murphy, the subject line of which read, "ISMRD Welcomes ML III Families", opened up a whole new experience for many of us. Wow! That was the first time anyone welcomed us using the name of Kelley's disease! Paul explained that, as the president of the International Society for Mannosidosis & Related Diseases (ISMRD), he received my address from Jenny Noble of New Zealand. I had previously corresponded with Jenny, since the Nobles are the parents of three children, two of whom have ML III.

Paul and Deborah Murphy are parents of a daughter with a condition called Alpha-Mannosidosis. They were instrumental in developing ISMRD to fill a void they perceived existed for affected families as well as for scientists and physicians with an interest in research and treatment modalities. Their group consisted of those who were interested in a number of rare diseases that are similar to each other, and fall under the Glycoproteinosis class of Lysosomal Storage Diseases. The conditions of ML II and ML III were a natural fit for the organization. Just as we had experienced in our family, not knowing anyone else with ML III when our daughter was diagnosed at the age of ten, Paul and Debora knew of no other families affected by Alpha-Mannosidosis when their daughter, Taryn, was diagnosed, also at the age of ten. It was via the internet that Paul acquired knowledge of Taryn's disease and eventually made connections with other families and key scientists in various parts of the world. He created a web site in April 1997, under the auspices of

NYU Medical School, called *Rare Genetic Diseases in Children: An Internet Resource Gateway*, and the decision to form ISMRD resulted from a meeting between Paul and others involved with Mannosidosis in Baltimore, Maryland during the fall of 1998.

Upon registering as a Maryland corporation in March 1999, ISMRD began formulating its mission, which the founding Board of Directors (Paul & Debora Murphy, Melissa Feliciano and Doug Kriss) developed into a plan to advocate for families and caregivers of all children affected by a Glycoprotein & Storage Disorder. At that time these diseases encompassed Alpha-Mannosidosis, Aspartylglucosaminuria, Beta-Mannosidosis, Fucosidosis, Galactosialidosis, Schindler Disease and Sialidosis. However, as scientists began including the Mucolipidoses in the Glycoproteinoses class, it seemed fitting to open the umbrella to include these diseases. The first official act for ISMRD was to send its President, Paul Murphy, to the 5th International Conference on MPS & Related Diseases in Vienna, Austria in March of 1999. It was there that the fifth original member of ISMRD's Board, John Forman of New Zealand, was recruited.

It readily became apparent that ISMRD should help promote cooperation and advocacy efforts with the other Lysosomal Disease organizations worldwide. Two of ISMRD's directors, Paul Murphy and John Forman, were instrumental in coalescing the global cooperative network for Lysosomal Diseases that was soon realized through organizations such as the Global Organization for Lysosomal Diseases.

A key objective for ISMRD's Board was to gain formal recognition of Glycoprotein Storage Diseases in the research community, where there were no cohesive activities being pursued at that time. With that goal in mind, ISMRD approached the Director of the Office of Rare Diseases (ORD) in Washington, D.C., in the fall of 1999 to initiate the process for convening a focused scientific workshop, funded through the National Institutes of Health. The support for this goal was immediate and enthusiastic and the long, slow process to build the case for the efficacy of such a scientific conference began.

By the winter of 2003, the National Institute for Neurological Diseases and Stroke (NINDS) formally accepted the goals of this scientific workshop by signing on as principal funder and organizer. At that same time, ISMRD's Board expanded the organization's disease coverage to two additional diseases that became part of the focus of the scientific workshop: Mucolipidosis II and Mucolipidosis III, and began planning a concurrent family conference for families affected by a Glycoprotein Storage Disease. In 2004, ISMRD successfully achieved formal status as a 501 (c) (3) organization upon review of its first five years of existence by the IRS.

Paul's initial contact with me was to invite us and other ML families to the workshop being planned for April of 2004. I wasted no time in sending email messages and making phone calls to the families I had contacted previously. They, in turn, contacted other ML families and the number of participants in the organization started to grow.

Although we were not able to attend that particular conference, as Kelley was due to have knee replacement surgery, the World's First Scientific and Family Conference on Glycoprotein Storage Diseases took place in April 2004 with some of the ML families attending. It was most beneficial for them to be able to meet each other, as well as the members of ISMRD who shared similar rare conditions. Approximately 25 families from the US, Canada, New Zealand, Australia, England, France, Norway and Latvia navigated cultural and language barriers to bond and become empowered with knowledge about the diseases affecting their loved ones. The conference resulted in three new volunteers for ISMRD's Board and a renewed commitment to press forward for continuing progress with its mission. The participants shared a vision of the future with the goal to identify affected patients wherever they may reside; to document the varied clinical, psychological and social implications of each disease and to support families, researchers and physicians.

In April of 2004, ISMRD provided all members with an on-line forum, the *Penguin Café*, in order for us to meet easily and share our stories as well as ask questions of each other. The founders of ISMRD had chosen penguins as a symbol, due to the fact that penguins characterize both the important bond of family, especially between parent and child, and the belief that the impossible can be overcome. To understand the nature of penguins in real life is to understand the correlation to ISMRD, its values and the depth of the bond within families affected by a Glycoprotein Storage Disease. Penguins are flightless birds that have adapted over time to flightlessness by evolving their bodies to a form that has enabled them to become masters of the ocean. Most penguin species have developed patterns of behavior that allow them to adapt to the severity of the climates in which they reside. Penguin mates share completely in raising their chicks and have one of the most unusual breeding strategies of all birds. Breeding is usually done in colonies. After the egg has been laid by the mother, the father incubates the egg between his feet and the folds of his belly. The mother will then go out to the ocean for up to two months to fish and feed, while the father stays back to maintain the incubation process. During this time, the father penguin will not eat, while huddled together in large masses to withstand severely cold temperatures and preserve body heat. After hatching, the mother returns and relieves the father of duty, so that he can travel an enormous distance back to sea to finally feed.

Like real penguins, ISMRD "penguins" show great adaptability to their harsh environment, manifested in the diseases that have afflicted their children. They are learning to symbolically fly and overcome the disadvantages of genetics and the clinical symptoms that have ensued. These penguins will eventually 'fly' by finding treatments and cures for these diseases and, in the meantime, will find ways to support one another across nations and differences in language.

Along with other families, we started off in the Penguin Café by introducing ourselves, our families and our geographical locations. I wrote a letter to the MPS Courage Magazine to let anyone else with any of the rare diseases focused on by ISMRD know that new members were welcome.

The Penguin Café rapidly became a vehicle for sharing our mutual experiences while dealing with the daily ups and downs that come with coping with a rare disorder. I think for most of us, it was simply nice to know we were not alone in trying to stay positive, despite the many trying situations we were facing, and that there was a place to go when we needed to blow off steam or ask for advice from someone else who understands. On a personal level, I was excited every time I saw that we had added another member, after all of those years of never knowing anyone else with ML III! We had much to learn from each other as we connected worldwide.

Autumn Tobey

Autumn's mother, Pam, joined the MPS Society after she learned about it. Since there were only a few ML III families listed in the directory in 2001, Pam and I started to correspond. We shared the feeling of being very isolated and alone when the doctors told us that our daughters were the only patients they had ever seen with this rare disease. While they could try to treat the problems, the doctors also wanted more information as to what course this disorder would take next. We found it to be a relief to be able to correspond with someone with whom we could share our feelings; someone who could understand what we were talking about.

We spent the next few years sharing many of our experiences via email, comparing notes and giving each other moral support as our daughters went through their trials. When others joined the MPS Society, they were included in our correspondence. However, we all felt a little lost within the MPS Society, as there was very little mention of ML – it was simply mentioned under the title of "Related Diseases."

In 2003, I passed along the information to Pam that ML II and ML III families were invited to join up with the families in ISMRD. She not only joined, but she enthusiastically became involved. Pam was grateful to have a place where she felt like her family really belonged. She saw ISMRD as being a vehicle for finding a way of contributing toward bringing people together who had the same goals in mind. Her willingness to do anything she could to further the work of the group, traveling to the conferences, and tirelessly working for the group endeavors, eventually led to her becoming a member of the ISMRD Board. In an effort to do everything she can to bring about awareness of ML, she has conducted numerous fundraising activities, all while running her own business.

Pam made a CD of beautiful spiritual music to sell, with the proceeds of those sales going toward research. She has sold jewelry, cookies, and plants, as just some of her fundraising activities. Of equal importance, Pam has consistently reached out to other parents, giving them both empathy and hope. She knows what it is like to have spent many years when absolutely no one knew what her family was dealing with, and she hopes that no one else will ever have to go through the same trials alone.

Allison Dennis

Trish was delighted that, when searching the internet, she found the Penguin Café site that is shared by the members of ISMRD and she introduced herself to the group in July of 2008. She explained that they had been feeling quite isolated before being able to communicate with others who were going through similar problems. They did feel fortunate to have met a few others with similar problems at an MPS Conference a few years prior to that. When the ISMRD group set up a link on Facebook, both Trish and her daughter, Alli, joined that group. They became frequent contributors, providing both information and entertainment to others who are also dealing with various aspects of ML and similar diseases. Trish states that she finds all of the ML sufferers display strength, determination, resilience and endurance.

Alli's thirtieth birthday was shared with many family members and good friends travelling from far away to join them. It was also shared with all of us, thanks to the internet. Afterwards, when Trish was looking through the amazing photos of her precious daughter, she thought of how truly blessed she is that so many wonderful new friends have come into their lives to help them and share this journey with them.

And, when they held a fundraiser to support research, they knew for sure that they no longer need to feel alone or isolated. For them the fundraiser was an extraordinary experience, where they felt much love from those who shared the special event that they had been told many times would not happen.

Alli became somewhat of a celebrity when she was interviewed for a television program in Australia, including a wonderful day full of surprises. Even though it is really hard to deal with such a disease, knowing that they are no longer all alone makes it all so much easier for the Dennis family.

Hayden and Sarah Noble

The first time Jenny met other ML families was in 1999, when they traveled from New Zealand to an International MPS and Related Diseases meeting in Manchester, UK. There were nine families in attendance. That is where she first met Dr. Jules Leroy, Prof. David Sillence and Ed Wraith. These three men have played a huge role in her life, while she has certainly challenged them to think about the real issues for ML patients.

That first connection meant everything to the Noble family. Jenny had come through postnatal depression plus the shock of being told they had two children with a life threatening disease. Little did she realize, at the time, that this was going to be a special journey that would take her from the very early beginnings of trying to understand what was happening to her children, to being the person who has challenged many medical minds, to being a co-author of a paper for Pamidronate that has gone on to help many Lysosomal Storage Diseases patients all around the world. That first meeting cost them dearly but they would never regret the decision to go. They had taken a second mortgage out on their house so that they could meet with a doctor in Maryland who first described the type of spinal surgery Hayden needed. From there they travelled to the UK to the Conference. Jenny remembers walking into the hotel and seeing all the families with different disorders; to her it was like going home. The ML families were able to spend two hours at their workshop with Dr. Leroy and the CEO of the UK MPS Society. From that workshop Dr. Leroy secured a room for all of them, and he spent time with each family. He was very giving of his time, and they returned home with valuable information. The Nobles have kept in touch with many of the families they met several times when they were overseas. That bond has endured through time, and Jenny has continued to gain the strength to handle the many challenges her children have.

In 2005, when the Nobles ran into problems with securing the proper health care for Hayden and Sarah due to the complexity of their conditions, they realized that it sometimes requires someone from the outside to step in and help.   Although Jenny wishes that part of their journey never had to happen, she is most grateful that when they reached out to Lysosomal Diseases New Zealand for help, John Forman (chairperson for LDNZ) became their advocate and helped them through the difficult issues of appropriate access to services at the Hospital, and also for specialist services across District Health board boundaries.

Jenny's devotion to her own children has been extended to many other families who are dealing with these rare debilitating diseases. She has spent countless hours working with others on a one-to-one basis, as well as to prepare and present conferences.

Since 1999 Jenny has been the Field Officer for Lysosomal Diseases New Zealand, where she is one of the founding members. She has worked alongside Prof. David Sillence since 2000 to help others learn about the benefits of the drug Pamidronate. She is the Project Manager for the New Zealand Organization for Rare Disorders, and was on the Board of Directors for the Australian MPS Society. She is currently Vice President for ISMRD, having been a board member since 2004.

Huddy and Sammy Anthony

The boys were diagnosed in with ML II 1993, when the Anthony family lived in Wisconsin. Since the time of that diagnosis, Liz has learned to cultivate gratitude for the positives she finds in their lives. Tom's parents were very available to take care of the boys on a weekly basis so Liz could attend a women's Bible study group. They even babysat to allow Tom and Liz to take two separate extended trips for much-needed respite.

In addition to grandparents, the family also had the support of aunts, uncles and cousins at large gatherings for holidays and birthdays. Everyone gave them great encouragement, and Huddy and Sammy were always pretty content and comfortable.

Their church family has always been friendly and available. The Anthonys have always been thankful for stable times of health because the smallest cold can be tantamount.

After they moved to Idaho in 2008, Liz, Tom and their children missed all of the interaction they previously had with family members, but they do manage to keep in touch on Facebook and the phone. They had caregiving in Idaho for one year and made a special friend who considers them family. They have been happy to be able to use the internet to keep in touch with other ML families. The few ML II children that they met at a Minneapolis conference for the MPS Society in 1995 have already gone to heaven. Liz has felt grateful that she has actually been able to meet MPS parents in Idaho, quite by accident.

Although there are some relatives and friends who do not seem to know how to relate to their family and the problems they face, Liz does not become bothered by it, believing that her energy is needed by her family. She has found the ML and MPS community to be an exceptional group for sharing, encouraging, and praying for one another.

Callie Nagle

Debbie is grateful to have had some great support throughout the years from their family and some of their friends at church, but she has also felt that the support she has received from other ML families has been invaluable.

After learning about Callie's diagnosis, Debbie found the MPS Society and the family joined, so when ML families were invited to join with the families in ISMRD, the Nagles were on the contact list. They were very happy to join and to attend conferences whenever possible. Debbie feels that there really aren't any words to truly describe how comforting it is to have all of the other families in their lives; the biggest connection is in finding people who you do fit in with, because this is a world where we don't always fit in. She has found the ISMRD internet site to be a blessing, and prays every day to God for the people who have been put on their path - especially friends who are accepting of Callie. She sees it as a beautiful thing that we all know each other's pain so well and none of us have to try to imagine what it's like. We have walked in each other's shoes!

Debbie likes being able to check with others to see if some of Callie's problems are related to ML or not. She knows that she can ask a question and receive many answers, not only as to the identification of the problem, but the way in which other parents were dealing with it. Prior to making the connection with other ML families, there was no one that understood or could relate in the same way.

Debbie has been very open in sharing her thoughts, feelings, encouragements, fears and prayers with other families. She is happy to have found a place where we can be here for one another and help each other when we're down; a place where we are all equally important. Debbie thanks God for such blessings!

Jennifer Klein

My first correspondence with Jennifer's mother, Terri, was in 2001 as we connected via email through the MPS Society. She was just coming to grips with Jennifer's true diagnosis and feeling relieved to know what was causing her daughter's mounting medical issues, while also feeling very anxious, since very little was known about the disease. Terri had participated with another support group when it was believed that Jennifer had a condition called Spono epiphyseal Dysplasia and was feeling in need of connection with someone who understood what she was dealing with, after having lost both of her loving parents in the years preceding this shocking news. We shared many messages comparing the way in which the disease affected our daughters and the attributes they had in common, such as being strong willed, independent and positive.

When Terri learned about the invitation to join the ISMRD, she enthusiastically became immersed, introducing herself and her family in the Café in June 2004, after which she was quick to give positive feedback to other members. By November of that year she had become so involved with the members that she was invited to serve on the Board of Directors of ISMRD. While in that position, Terri worked tirelessly with the other members to put on a fundraising Walk-Run to support ISMRD, an endeavor that evolved into a Conference with families and professionals as well as the fundraiser, bringing together over one hundred people in Ann Arbor Michigan in April 2005. We traveled from all around the United States and as far away as New Zealand and Italy to attend *"Crossing Oceans for a Cure."*

The Conference proved to be a wonderful opportunity for the participants to meet each other, share their stories, and gain knowledge from the doctors who presented valuable information that the parents would have had to travel far and wide to obtain.

The Walk/Run fundraiser followed the Conference, with most of the families participating, as well as many residents of the Ann Arbor area. Everyone involved agreed that it was an amazing experience. The fundraiser was a success, despite the fact that the weather was a bit unpleasant. The children from Florida who attended didn't think the weather was all that unpleasant, since they had never seen snow before!

By the end of 2005 the Board of Directors of ISMRD decided there was enough money in the treasury to appointment of Terri Klein as ISMRD's Executive Director, a position she took in January 2006. During the years that Terri held that position she worked to advocate for the Glycoprotein Storage

Disease family, focusing on making ISMRD more effective, and giving the group a more visible presence among the global Lysosomal Disease community. She travelled to many conferences and made numerous connections with physicians who had interest in Lysosomal Storage Diseases. She held more fundraisers and conferences, met with families, worked with other Lysosomal groups, and promoted the needs of the families while speaking in numerous public venues and seeking ways to bring about awareness and raise additional funds for research.

In 2009, Terri accepted the position of Development Director for the MPS Society where she continues to work for all of the MPS and ML families. Terri has not only found support for her own family, but has been available to support many other families, as she also retains her membership in the ISMRD.

Andre Andrews

For many years after Jane received Andre's diagnosis, she attended MPS Conferences where she felt very frustrated while waiting to hear something said about the disease her son had. She wondered if they remembered that ML was a part of the syndromes covered by the Society. After talking to one of the MPS parents about the oversight, that parent approached one of the doctors on her behalf. He listened and very nicely prepared a slide for ML right on the spot so that she could feel like she was a part of the conference as well.

Then, in 2004 she was able to attend the Conference that was held by ISMRD in Maryland. From her first day there, she found a sense of belonging from the first day that continued to grow through the forum of the Penguin Café far beyond her wildest expectations, finding everyone to be very helpful. She became a frequent contributor to the Café while getting to know the other families and comparing situations in their lives.

Whenever possible, Jane has continued to meet with other families, feeling that each encounter is a wonderful experience where nothing could beat the one-on-one encounter with another ML parent. The encounters have been like family reunions to her, which are all too short, leaving everyone wishing they could last much longer. At these gatherings, Jane has found a lot of joy, not only with the child and parents, but with the siblings as well.

Being so involved in daily life with Andre's school, before and after care, work, bowling and other legal issues that she has to take care of, leaves her with barely enough time to do anything else. So, she is very grateful to

Paul Murphy and the ISMRD for making it possible to sometimes just get away and spend time with the other families. It makes all the difference in the world to her and Andre. She sees the group as a special gift - an extended family that she can count on.

Joey Nagy

It was shortly after she learned that her son Joey has ML III that Linda logged onto the ISMRD website in 2004 and told the group members that she was very excited to find them. Although it was a relief to finally have a name for what is happening to Joey, Linda was terrified by not knowing what to expect. She found it hard to comprehend exactly how rare the disease is, and was grateful that she could correspond with other parents.

As Linda became involved she wanted to know what she could do, as well as what the group could do, to try to bring about awareness. She found the parents of these special, wonderful children to be an awesome group, who she believed God chose because He knew that they would love and take care of them, and fight for them, giving 100 percent of their energy. Joey's mother didn't want to accept phrases like "I can't" or "It cannot be done" and wanted to know what she could do to help the group to make changes that would bring about awareness, even though it was still new to their family.

Linda posted that when she first heard the word Mucolipidosis, it was the day their lives changed forever. And, although it saddened her to no end, she was relieved to find the silver lining when she thought that there wasn't even a glimmer of hope. ISMRD and her faith in God changed all that. She found it comforting to know that she had somewhere to go where she would not be judged for expressing her feelings.

With hopes that dealing with the news of her son's disease would become easier as time went on, Linda was very excited to be able to attend a conference and meet with other families. She felt that the parents she met through the group were the most awesome, loving, strong parents she had known. They showed her that there is a lot of living after diagnosis. Linda valued the advice she received and the connections she was able to make.

Returning home after their first Conference attendance, Linda felt a bit lost without their new-found family. She was amazed at how close everyone had become. Joey missed his special friends, so Linda and Frank started to reach out to others, holding gatherings in their area so those who could attend would be able to spend precious moments together just sharing and having some fun.

While working on acceptance, she felt strongly that one of the biggest hopes for the children is not only early diagnosis but also, sadly, finding others and increasing the numbers in the group. She wondered how many others there might be like Joey, who went nine years without a diagnosis, and if the disease is quite as rare as it appears to be. As often as possible, Linda has made every effort to be in touch with and offer support to those who are traveling down the same path that they are.

Anna James

Shortly after learning about Anna's diagnosis, Jackie and Brent joined the MPS Society, and Jackie and I started to correspond with each other via email. Being anxious to meet someone else who was dealing with the same disease as her daughter, Jackie talked Brent into taking their vacation in New England… and slipping in a visit with our family in New Hampshire. In August 2002 Jackie and Bret came for lunch with their two children, eleven-year-old Peter and seven-year-old Anna. We had not exchanged photos before our meeting, so none of us had any idea what to expect. As it happened, Anna sat next to Kelley at the table. Bob and I kept looking at Anna, and then exchanging glances with each other. She looked so much like Kelley did some thirty-plus years earlier; we couldn't quite believe our eyes. Kelley later remarked that she had never expected to see anyone else that looked like her. Little Anna had the same type of hands, arms, legs, impish smile, eyeglasses and just plain cuteness that Kelley had at that age.

We had a wonderful visit getting to know the James family and promised to see them at the next MPS Convention, in their city of St. Louis. We did just that in April 2003. It was the first time that we saw another ML III family at one of the MPS Conventions.

Jackie was quick to join up with the ISMRD once the invitation was extended to ML families, and she started to connect with all of the other families, sharing information and attending every conference and family gathering that she could. She started to post on the Penguin Café where she found it to be amazing to be in touch with others that are experiencing the same issues and could really understand what their family was going through.

She incorporated fund-raising activities into her life and her business in St. Louis, helping in every way she could to raise awareness of the rare disease of Mucolipidosis. And, in 2011, she accepted a position on the Board of Directors of ISMRD, where she applies her energy and know-how to raise funds to apply to research.

Spencer Gates

Spencer's mother, Andrea, is very thankful that there is a support group available for them. She has found the common bond shared by the families of the ISMRD to be a way to feel at home and among instant friends. She quickly became involved with other parents, being very excited for Spencer to see other kids with ML so he doesn't have to feel that he is the only one who is different. She also believes it has been good for their daughter, since when Sydney was young she used to ask if a person who appeared to be special had the same disease as Spencer or something else.

The Gates family has attended every conference and simple get-together with others whenever it has been possible. In 2009 they put on a large charity dinner and auction to raise money for ISMRD.

Andrea feels that, with ISMRD being her connection to people who truly know what her family goes through; she has found a place that is making a difference in Spencer's life. When the Board members of ISMRD asked her if she would like to become a board member, she quickly accepted the position, and she has worked since then to help ISMRD to grow.

They are also involved with the MPS Society, participating in informational videos produced by them.

Sergio Cardenas

The first ML connection that Maria Elena made after searching for information on the internet was with Jenny Noble. It was shortly after they received the diagnosis for Sergio. After that, Maria Elena introduced herself on the ISMRD Penguin Café in 2004. They were still living in Venezuela then, and she didn't think her English was good enough at that time to be able to correspond as well as she wanted to. However, she was working to learn English, as they were planning to move to the U.S. that year, after waiting for two years to do so.

She had also connected with Brenda, whose son Zachie was just about the same age as Sergio, through the MPS on-line forum, where there were not many ML families. Maria Elena was very relieved to be able to communicate with other parents, via email and the Penguin Cafe, and to finally find someone else that felt the same way she was feeling, and living the same life on a 24/7 basis for 365 days a year.

Still, the language barrier felt like a big wall to her. Even though she took English classes for six years before she really needed to speak English, she never had the opportunity to practice it and know if she really had learned something since she lived in a Spanish speaking country. Yet, she learned how to chat in the Penguin Cafe and then met all of the members there. She felt like she finally found a family from all around the world. Maria Elena and Brenda Haggett learned that their sons, Sergio and Zachie, had much in common. Maria Elena already knew that Sergio had ML II/III, but nobody else talked about that type. Brenda started feeling the same way about Zachie, and then finally, when the doctors gained a better understanding of the disease, Zachie and a few other children were diagnosed as having the ML II/III type.

It was Brenda who talked with many people about Sergio and the Cardenas family to help find a good doctor for him, and found a doctor who has been Sergio's primary physician since they moved to Texas. Maria Elena will be thankful to Brenda forever! She feels very connected to her family, as well as all of the ML Penguin families, being relieved to not feel so alone with this horrible disease. She has made an enormous effort to learn more English words, to understand, to write, to read and to try to communicate, but it is difficult when one is used to using words in their native tongue to express their feelings and ask some very important questions. She has attended as many conferences and gatherings as she has been able to, finding it very helpful for her to be with other families, and for Sergio to be with the other ML children.

Maria Elena really misses her family in Venezuela, but since it is so expensive to travel there she tries not to think about it too much. She likes taking care of Sergio, the rest of the family, and everything at home, but some days she just needs to vent, and express her feelings. At those times she feels lucky, happy and blessed that she has other ML parents to talk to. Of course, she also hates the fact that other parents need to face ML, and she hates when other children need to fight against this monster disease, but she really likes to meet new families. She finds the Penguin Cafe more helpful than Facebook, because it is more private, and the themes stay there in files that she can go back and check for advice. However, she does acknowledge that people do love new things, and she also loves to stay up with current trends.

Sometimes Maria Elena feels sad that there are more ML II and ML III families, than there are MLII/III families, so there are only a few who have the same symptoms as Sergio. But in the end, she loves all the ML families, because all are in the same boat, and nobody judges her path or her steps, because they wear "the same pair of shoes" she wears every day.

Zachie Haggett

The Haggetts attended their first MPS Society Conference in 2004. It was exciting for John and Brenda, in that they were able to connect with other families, and exciting for Zachie who was thrilled to visit Disney World.

Even more exciting was the trip they took to Ann Arbor in April 2005 for the *Crossing Oceans for a Cure Conference* and Walk/Run Fundraiser. They saw it as a once in a lifetime opportunity to meet the two most knowledgeable ML doctors in the world and the largest gathering of ML families ever. To be able to speak to a doctor who has even heard of this disease, let alone discovered one, was beyond comparable. To meet in person more than twenty different people living with ML from all over the globe was beyond belief!

Almost two years after diagnosis, they were not feeling as alone anymore. Knowing they had a long road ahead and many battles to fight, they were thankful to have connections through the internet so they could stay in touch with everyone they had met.

Brenda became a frequent contributor to the Penguin Café and other internet sites where she shares her thoughts with others. The family has attended every gathering that they have been able to find. Connecting with other families who deal with MPS and ML daily is inspiring to them because they see many families as being very resilient. They gain courage to get through whatever they have to face, and share the mourning of the losses of those who have succumbed to these awful diseases.

In 2012, Zachie's family received the support of the Make-A-Wish Foundation, even though Brenda had been putting it off for years. Zachie's special service coordinator, who helps him get the services he needs, actually nominated Zachie. Brenda felt that he deserved his wish, but lamented that it was another part of the journey she had hoped would never have to become a reality. Still, she was grateful to Zachie's coordinator and doctors who gave their blessings, as well as everyone at Make-A-Wish and the organizations that sponsor a child to afford their dream, for making Zachie's dream come true.

Brenda felt that seeing Zachie's face light up when he learned that he would be going to Universal and Disney World, and staying at Give Kids the World Village, a nonprofit resort for children with life-threatening illnesses and their families, was worth more to them than any trip could ever be. He asked how soon he could "pack his luggage."

They were able to do everything Zachie wanted to, and he loved every single minute! His parents were thrilled to see him get excited about everything and very appreciative of everyone involved in making his wish a possibility.

It's not common to meet complete strangers who truly understand what a family is going through, and the Haggetts believe they have been blessed by many throughout the whole process. It left an enormous imprint on each of their hearts and most especially their small but mighty little survivor, Zachie. This family really appreciates everyone who shows them their love and kindness and understanding of life's true meaning.

There are no guarantees in the medical profession, and there are many schools of thought on any one subject. ML is such a rare disease, that all of the aspects of it are still not known. Not everyone with ML will have a severe bone problem, but some do. Not everyone with ML will have the Carpal Tunnel compression, but many do. Not everyone with ML will have vision problems, but some do. But all of them seem to have a great smile!

I look forward to the day when more is known in order to help ML patients, and I am so glad we have a place to stick together to try to get the recognition we need for our families.

CHAPTER FIVE

MEDICAL APPLIANCES FOR ALL OCCASIONS

In order to keep their children comfortable, out of hospitals and alive, families use a wide variety of appliances and equipment. From the simplicity of crutches to the intricacies of IV infusions, there are numerous ways in which some families have had to rearrange their homes and lives while caring for the special needs of their children. Although some medical insurances help with the cost, there are many small details of daily living that are added expenses to the family. In addition, there are expensive modifications that must be made to the homes of people with disabilities that are not covered by any insurance plans or government programs. It isn't possible to cover all of the adaptations we have made for our children, or the creative ways some people have found to meet their needs. Most of us have adopted a 'whatever works' attitude and these pages will give the reader an idea of the complexities added to our lives.

Kelley Crompton

Kelley's first encounters were with splints and braces. Next, she was placed in traction in the hospital, and then at home. Various casts and splints followed, with Kelley becoming proficient in the use of crutches at a young age. In addition to the use of crutches, Kelley occasionally used a wheelchair.

By the time she was sixteen, the kyphosis (abnormal curve) of her spine became severe. Her orthopedic doctor told her if she wore a Milwaukee Brace for a year, she probably would be able to avoid surgery to correct that problem. The brace was very large and uncomfortable. It was worn over a t-shirt, but under her regular clothes. Thus, her clothes didn't fit well. On the day she received the brace, we stopped on our way home to shop for new clothes. She was only allowed to take the brace off for an hour a day. It was a long year for her but that was one operation she was able to avoid.

After one of her hip and bone operations, she was placed in a large striker-cast that extended from her underarms to her ankle on the right side. The 'big clothes' she wore when she had the large brace came in handy again for the many months she wore the cast. In order for them to make the cast, she had to be totally nude while three men applied the material. She later remarked that, although the men talked about other things while applying it that was the day she lost all modesty. During the time she wore it, she had to use crutches to get around but she managed to devise many ways to return to a semblance of normality.

Kelley's breathing problems necessitated numerous different types of equipment, such as a suction machine, a nebulizer and supplemental oxygen. At times she had to use IV infusions to fight infections. Some of that equipment and medicine took up a lot of room in our house. Her visiting nurse taught me how to handle the various kinds of medications – sometimes three different interchangeable ones. When one of the medications required refrigeration, it was delivered to our home packed in dry ice, and it took up an entire shelf in our refrigerator.

During the last years of Kelley's life, we had many deliveries from the medical supply companies. We needed to use a suction machine, a feeding tube, supplemental oxygen, and many different medications. Keeping track of all of those medications was quite a project, so I created a chart. My sister suggested that we could use a vacuum shoot from the drug store to our home. The staff at our local drug store was very good to us, helping us handle the many different requirements imposed by insurance companies. They often asked me how Kelley was doing when I was in there to pick up medications, and they put up with me when I was having a bad day and losing patience with bureaucracy.

When we took her to the medical center for her various appointments with doctors and Pamidronate infusions, the amount of equipment I had to take with me required two large bags. I used one of the hospital wheelchairs to cart the stuff around when we were there.

In addition to all of those medical requirements, Kelley had a bad reaction to a medication, affecting her ears and disabling her sense of balance. She was left with a residual constant ringing in her ears (tinnitus). She requested another addition to the accumulation of her necessities. In order to take her mind off the tinnitus that was very distracting when she wanted to sleep, she thought that the sound of a small water fountain would help. The first one we tried did help, but it didn't last long, so we moved on to another that came with very explicit instructions for weekly cleaning. Because of the little light it contained (which also had to be replaced frequently), that became a delicate procedure. When that one failed, we moved along to another. All in all, we wore out many different fountains during those years. And today, whenever I am somewhere where there is a fountain, I think of how much Kelley would have appreciated that sound.

Autumn Tobey

Since most of Autumn's problems were with her bones and joints, she employed the assistance of a variety of special appliances, including many different walking casts and night braces that were constructed specifically to help straighten her fingers. She was still misdiagnosed with JRA when her fingers started to contract at the age of six, so the doctors advised intensive Physical Therapy, which took place three times a day. She had a hot wax bath for her wrists, hands and fingers followed by the manipulations that Pam had been taught to do. Autumn became very tired of that routine. The braces that she wore at night had her fingers hanging from trapeze type bars on top of the brace. Pam said that they looked funky… resembling the hands of Edward Scissorhands… or perhaps Freddie Kruger. Pam didn't know how the child slept!

With the passing of years, Autumn has undergone many more surgeries, after which she has had to do more extensive and painful Physical Therapy and use various splints, braces and crutches. When walking distances became difficult, they found it necessary to obtain a wheelchair. Wanting to remain as independent as possible, Autumn didn't want the wheelchair at first, but she did accept it once she realized how much it really has helped her.

Allison Dennis

Alli always had issues with walking; even as a toddler she would simply sit down and refuse to walk. Her parents, Trish and Richard, thought it was just a case of stubbornness, never realizing what the real problem was. By the time she reached the age of ten, Alli didn't want to go anywhere – indeed, begging to stay home. When she was finally diagnosed with ML III, they were told that she would probably need a wheelchair in the future. They tried not to focus on that.

By the time Alli was eleven years old, Trish decided they needed to get a wheelchair so she could start to enjoy her life again. However, she had no idea where to start looking. At that time they had no coordinating services, case worker, or anyone to ask for help, so she searched the local classifieds. Through her research she found a non-profit organization that provides support and services to people in AU with a broad range of disabilities. After waiting six months for the appointment, they were able to use a loaned wheelchair for Alli. It started to free Alli from some of the restrictions to her mobility. Next they had to make an appointment with the Occupational Therapist to get a custom chair made. Alli was twelve when she finally received her first custom chair. Over the years, she has had five manual and four electric wheelchairs. As she has grown and her condition has changed, chairs became unsuitable and a change was necessary.

Alli was twenty-three when she started using a wheelchair on a full time basis, after she had a hip operation in 2006. By 2011, her pain level being high, she required an electric wheelchair that has tilt, space, recline, and up and down features. She has found it to be an excellent assist in taking pressure off of her spine, as she can no longer sit for extended periods of time. This chair has specialized seating and back rests; it has helped to give her some feeling of independence. Alli also has a shower commode, a hoist medical bed and an alternating air mattress. These all help with her daily care. Despite all of this, Alli has reluctantly become very dependent on her mother since she can no longer do very much at all for herself. This is difficult for her as she has always been as independent as she could and hates to have to accept help.

In 2013 the family bought their first (second-hand) wheelchair car. They were pleased that it was in awesome condition, easy to drive and very easy to strap Alli's wheelchair into. It has made their lives easier with less pain for Alli when they go out, as she can be transported in her wheelchair with no need to transfer. Trish's back is also feeling less of a strain now that she is not lifting Alli to transfer her from her wheelchair into the car.

Hayden and Sarah Noble

At the age of nine, Hayden used a stroller to travel long distances. A few years later a wheelchair was required, but only for long distance travel. Yet, by the time he was fifteen, he became paraplegic. Since that time, he has used a wheelchair full-time.

Both Hayden and Sarah have used many different types of braces over the years, including night-time hand splints that didn't work very well. Hayden also trialed ankle braces and then leg braces at night to keep his legs straight, but those didn't work either.

Sarah was five years old when she used a stroller for long distances only. She required the use of a wheelchair full-time at the age of thirteen. Since the age of fifteen, she has needed the wheelchair for long distances only and uses a walking frame in her home. She also has a bed board which helps her turn over in bed.

Both Hayden and Sarah use a shower chair for showering. Due to the limited range of motion they both have, they use shower brushes that help with washing hair and scrubbing their backs, legs and feet. They use a shower trolley for soap and shampoo, etc. And, as with other ML patients, many adaptations have been made to their home to meet the needs of both Hayden and Sarah.

Huddy and Sammy Anthony

Tom and Liz Anthony consider themselves to be ordinary parents. Yet, the accommodations they have made to take care of their ML II boys make their home anything but ordinary. Both Huddy, born in 1989, and Sammy, born in 1991, have needed Physical Therapy, Occupational Therapy, Speech Therapy and Water Therapy spanning decades. They have used numerous double strollers, two custom walkers, and a custom wheelchair.

Sammy and Huddy both stopped growing physically when they reached about four or five years of age, so neither boy ever weighed over fifty pounds or grew taller than three feet. The boys took numerous medications, wore diapers and used bed pads. The Anthonys created some commonsense solutions to make accommodations for the boys, like putting their mattress directly onto the bedroom floor when they were toddlers so they could crawl in and out of bed themselves. They bought various Little Tikes furniture, like a picnic table with chairs, for them to sit at for lunch and during their home-school time. Sammy and Huddy were always able to eat and drink a normal diet, but their food had to be cut up for them so they wouldn't choke. Since the boys crawled around all day, Liz and Tom were careful to keep the rug and floor extra clean.

Both Huddy and Sammy wore inflatable splints at night that were placed over their knees to prevent their contractures from becoming more severe. Shoe splints were used during the day to prevent ankle/foot contractures and toe-walking.

A Rifton Supine standing frame is an adjustable frame holding a board which provides the ability to stand, promoting therapeutic benefits, including head support with partial weight-bearing. The family had one of them sitting in their (then) nine-hundred-square-foot living room for several years, since Huddy used it almost daily for up to an hour each.

They also used therapy balls, a mini trampoline, soft helmets, and safety tricycles. Specialized seat belts were necessary, including a harness for lying prone in the car and a Y harness. In addition to glasses, they each needed hearing aids for both ears.

For breathing treatments, they used nebulizers, a pulse oximeter, oxygen concentrator and a large oxygen tank. To facilitate noninvasive airway ventilation, they also used CPAP (Continuous Positive Airway Pressure), BiPAP (Biphasic Positive Airway Pressure), and VPAP (Volume-Assured Pressure Support) machines, requiring tubing, masks, cannulas, and filters.

The cost of some of the equipment and supplies were covered under their primary insurance through Tom's work, and supplemented by a secondary insurance program (The Katie Beckett Program in WI and several other states), which subsidizes special needs children who qualify, in order to keep them living at home in the care of their parents.. However, there are numerous supplies that the Anthonys have paid for out-of-pocket because their children needed them. Although they have been able to use an average sized van with special seat belts, since the boys weighed only forty and fifty pounds each, theirs is not an ordinary home!

Callie Nagle

Due to many of her joints being affected by ML III, Callie has needed to use a variety of braces and splints throughout her life. Those include a neck brace, a back brace, knee braces, ankle braces and hand splints. The knee braces, prescribed by her orthopedist, are soft, and they attach with Velcro. They help keep her knee caps in line, because they otherwise float. She also wears orthotics in her shoes. All of these pieces together have helped. Physical Therapy is crucial as well - stretching those calf muscles! Callie finds it more comfortable to walk on the balls of her feet because her calf muscles are very tight.

Although she has found it necessary to use crutches at times in the past, she is a very strong willed girl, so it took her a long time to give in to the hip pain and use them full time. A wheelchair was finally needed for any outings while she was waiting for her hip replacement surgery. Debbie has tried to encourage her to use devices like grabbers and jar openers, but Callie has always been reluctant to admit she needs to.

Jennifer Klein

When she was nine years old, Jenny had to wear a body brace for six months after surgery in order to be sure the correction made to her spine would not be jeopardized. It was called a clamshell case. However, it was more like a turtle shell type of body brace for her upper torso, extending from her underarms to her hip bones. It was composed of plastic and had Velcro applied, so she could take it off for showering. There were breathing holes in it, and it was worn over a t-shirt. She wore it for six months, and became dependent on it to the extent that she felt panicky when she didn't have it on, and was anxious to be without it when she went back to school.

Jennifer has an amazing amount of upper body strength, which has served her very well the numerous times she has needed to use crutches, particularly when she was in the fifth through seventh grades. Although her arms and shoulders held out very well when she has had to walk long distances using crutches, her hands often became blistered, so her parents devised numerous ways of using memory foam for padding.

By the time she was ten years old, Jenny was having difficulty walking distances; she could walk only about ten to fifteen minutes, before she had to take a break. During that year Terri and Walt determined that she should use a wheelchair at least part-time. Jenny was willing to use it when the family went on trips, but she wasn't ready to use it at school quite yet. At the age of twelve, she admitted that getting around at school was quite difficult, so they acquired a 'passport' scooter for her. It has served her very well since then, although there still are times when she needs the assistance of a wheelchair.

Jenny's battery-operated scooter has a tight radius for turns within the home, and can travel up to ten mph when in such places as shopping centers. It has allowed her to be more independent and mobile, although she did require an aide at school to help open doors. The scooter can be broken down and reassembled very quickly. Now that Jenny is driving her own car and attending college, she can take the scooter with her to use when necessary.

Andre Andrews

Andre has some deformities in his wrists for which he wears splints at night. Contractures in his legs and arms causing tightness are minimized by physical therapy, occupational therapy, and pool therapy. For years, Andre slept with a CPAP (Continuous Positive Airway Pressure) before it was determined that a BiPAP (Biphasic Positive Airway Pressure) would better meet his needs to facilitate noninvasive airway ventilation.

When Jane and Andre took a long distance trip, his special car seat somehow became badly damaged while in the care of the airline employees. Jane assumed that it may just be a recurring situation, after they offered a replacement car seat and took them to a room full of new car seats. They were able to find one to accommodate Andre, although it was still not the one that had been recommended through physical medicine, since the insurance company seemed to think a specialized car seat is an over-the-counter product. Jane fought for one that would make Andre as comfortable as possible. She found it frustrating that they didn't understand how difficult it is for someone whose legs don't bend in the normal fashion and wondered why they didn't take such physical limitations taken into account.

For many years, Andre has been unable to walk due to unbearable leg pain. He keeps his legs in a semi bent position, which makes sitting in his chair and on his bed easy, but standing in the tub to bathe has been a real challenge. Yet, when Jane tried to get him a bath chair through the insurance company, they denied it as an accessory.

When Andre went to school, his mother had problems finding before and after care for him. No one wanted to provide services for a child who was unable to walk and who had to be pushed around in a stroller. During the year, the school utilized his wheelchair in the daytime but due to fact that their housing did not give him the access for it, it remained at school year round.

Upon finally finding a daycare program to accommodate Andre for the entire year at a facility where the management believed that every facility is supposed to offer a special needs placement by law, Jane was very relieved. She also found a clinic that provides equipment recommendations. They worked with her to assure that Andre had everything he needed. They were proficient in countering denials from her insurance company, as she experienced with her initial request for a bath chair and dinner chair. While she saw that as a positive step, it still bothered her to hear the term 'Special Needs' used in regard to her son. Yet, prayers were being answered when, all of a sudden, they were being directed to resources that no one ever told her were available for her child.

Andre has run into substantial issues with regard to handicap access. He had a one-on-one aide in school and will probably always need one-on-one assistance since some buildings and doors are not accessible for him due to the restrictions he has with reach. He is unable raise his arms high enough to reach elevator buttons or door access buttons, and not all doors have a mechanism to open them automatically. Jane has concerns as to how independent Andre will be able to be. She found it a real eye opening experience once she started letting him go out more in his wheelchair. He's unable to reach in his pockets, so things like paying for items with his own money is an issue. In order to use a credit card, Andre would have to be able to see the transactions, which led one of his therapists to suggest either raising his seat or providing an adjustment that would allow his seat to go up. The question then would be whether or not he would be able to reach that switch. Most tables are the same height as his wheelchair so he can ride to them and make himself comfortable, but he requires assistance with cutting food and moving things into the position of his reach so he can reach his cup as well as his fork. He will always require assistance dressing, going to the bathroom, and transferring to and from his wheelchair.

Andre now has what they consider to be an awesome wheelchair, although it has caused some challenges getting into Jane's van because it is larger than his previous chair, which was a perfect fit. There are doorway issues as well, but it is state of the art, with many accessories. It can be raised up high, which is a blessing for Jane's back! He can lower it all the way to the floor in the event he wants to watch television from a different viewpoint, and he can lean it back to relax. Best of all, it helps Andre to be as independent as possible.

Joey Nagy

At the onset of Joey's physical problems, his orthopedic doctor sent him to physical therapy where they made a variety of splints for him. He responded well to that treatment.

He was only eighteen-months-old when Joey saw an eye specialist because they were checking for clouding of the cornea, a symptom found with MPS conditions. While he didn't have the clouding, it was determined that he needed glasses at that tender age. His parents were concerned about how he would keep them on. The doctor reassured them that it shouldn't be a problem, because once a child figures out that they can see better, they actually ask for them. He was right. They never had a problem with Joey wearing them. But they did have a problem keeping him in a functional pair. The glasses posed a constant problem. Joey never had much of a nasal bridge, so they would constantly slide down a bit. He has always felt as if his glasses weren't where they were supposed to be. Whenever Joey became angry, he would rip off the glasses and send them sailing across the room. They usually became bent out of shape, but he always asked ask for them back immediately. In time, they learned to get a second pair every time his prescription changed, as well as a warranty for scratches to the lenses and damage to the frame. They became on a first name basis at the vision center, where the staff would say, "Here comes Joey."

By the time he was ten years old Joey was diagnosed with moderate to severe sleep apnea, a medical condition in which breathing ceases for a few seconds, or even minutes, during sleep. They determined that his airways were too small and recommended removal of his tonsils and adenoids. The procedure is usually done on an outpatient basis, but because of ML, they wanted him to spend the night. That surgery went well and further testing showed that his obstructive sleep apnea had tremendously improved from the removal of the tonsils and adenoids. However, he still does have mild Central Sleep Apnea and was referred to a neurologist. A previous neurologist had ordered an EEG (electroencephalogram) to determine Joey's brain activity and started him on seizure medicine because of an irregularity, even though he had never actually exhibited a seizure. Joey uses a CPAP (Continuous Positive Airway Pressure) machine every night, as the doctor believes it will help preserve his heart because it will not have to work as hard.

During his tenth year Joey's parents started thinking about acquiring a chair for him to use for long distances. They were no longer able to take family walks, because it was too difficult for him. If he went to play at the school with his brother, he always returned home feeling very sore. They didn't want to push him into something prematurely, but on the other hand, they didn't want him to be limited. It took them a while to make the difficult decision, but as Joey experienced increasing pain and he had more trouble getting around, his doctor believed it was time. He explained to Joey that he wouldn't have to use the chair all the time, but it would be there for when he needed it. They weren't sure if they should start with a manual chair or go right to a motor chair or scooter. After a demo of a scooter, they decided it was the perfect solution.

Anna James

Jackie and Bret found that it was difficult to know where to draw the line between being too soft on Anna, and forcing her to press on. As she grew, her mobility began to decline to the point where it severely affected their family's recreational opportunities. When they went out to the store she was able to sit in a cart but if they wanted to enjoy outings to parks, zoos, museums or take long walks, Anna couldn't join them, and they felt guilty about leaving her at home. As time progressed, her hip sockets basically eroded away and the use of a wheelchair on a full-time basis became a necessity. When they started pursuing the idea of getting a wheelchair for her, they found it to be a very frustrating situation. While many institutions offer wheelchairs for use, some (including their zoo) charge a rental fee. Jackie struggled with home health providers and their insurance company to obtain a chair for Anna. The equipment providers their insurance company approved either did not return calls or were unable to find a vendor that would offer a pediatric chair that met their needs.

The James family lives in a home that was built in the 1880's. When they bought the house, Anna was able to walk and climb the stairs. That lasted for about four years, after which it became necessary for Jackie to carry her from the car and up the several steps to the house. After catching her breath, she then proceeded to carry her up the twenty steps to her bedroom. As Anna and Jackie both became older, that became more difficult for Jackie to do. In searching for help, they talked to numerous agencies but were told that they earned too much money to qualify for any assistance. They found that to be very frustrating, since both parents were working very hard to provide for their family.

One day, when Jackie was attending a concert for students, she sat next to another family with a child in a wheelchair. That was when she learned about a local organization that help special needs families to make adaptations to their homes. Jackie called them immediately and had someone visit their house to learn what their needs were. They advised her that they determined what would be done based on need, not on finances. And they determined that, indeed, Anna was very much in need! They promised to install a stair-lift and a complete renovation to the upstairs bathroom. They also approved an outdoor ramp but, out of the blue, the family was approached by somebody in their church that has a ministry which builds wheelchair ramps for the needy. For two weekends, they had people they knew, as well as people they didn't know, show up and spend their precious free time building a fantastic ramp for Anna. It literally is like a deck! Anna was delighted to be able to wheel herself in and out of the house without needing to call anyone for help.

They expect that the local organization that has promised to give them the stair-lift will put in a remote controlled back door so that Anna can just press a button and get outside on her ramp.

The other wonderful help that Anna has received is a dog. This dog has been raised from the age of six weeks to help children and adults with special needs. Once again, a friend advised Jackie of this organization. Within a few months after they sent in all the required paperwork, they received a visit from a couple of people that interviewed them. In another few months they heard back from them that they had a dog named Tess that they thought would be perfect for Anna.

When Tess and her trainers visited, Anna and Tess seemed to take to each other very well. Bret and Jackie had to attend a class for four days to work with Tess and make sure they knew all the things that Tess could do for Anna. They had to learn certain words they need to use to tell Tess what they want her to do. They were also taught how to take care of her. Tess is not a fully certified service dog (their choice) but she does quite a few things for Anna. The most important is that she is a wonderful companion and friend for Anna. She can take Anna's glasses off, take Anna's socks off, close doors and drawers, pick things up from the floor and give them to Anna and she will take a note downstairs and give it to Jackie if Anna needs anything. They feel that Tess is one of the best things that ever happened for Anna and the entire family.

Spencer Gates

In 2004, Andrea started the process of obtaining a scooter for Spencer. A year later she was still getting a runaround. His doctor had the authorization within hours, but that's where it stopped. The company that had been approved to make the chair claimed they needed more information from the insurance company before they could even make an appointment. Everyone seemed to be blaming everyone else for the lack of action.

When the insurance company denied Spencer a custom wheelchair, Andrea and Kevin rented one. But it was not at all adequate for Spencer, who couldn't reach the plates. He was unable to wheel himself around, or even get into the chair by himself. Andrea spent hours making endless phone calls and waiting for doctor's letters in order to finally receive insurance approval for his custom wheelchair. Then Spencer's wheelchair order sat on someone's desk for two months, and the family had a change in insurance companies, so no one wanted to pay. It took a very long time to resolve those problems.

By 2010, after they had been living in a house where Andrea had to carry thirteen-year-old Spencer almost everywhere, the Gates family moved. They found a larger ranch style house with wider doorways, where Spencer is able to get around the whole house in a scooter, enabling him to be independent. They have a bathroom just for him, with a bath stool in his shower. His shower has a removable handle. The shower door allows easier access for him to get in and out, and the shower entrance has a plastic step to assist. The sink has a step so Spencer can climb and sit on the sink. He has a floating/air mattress along with a CPAP (Continuous Positive Airway Pressure) machine. Spencer uses an inside-walker and an outside-walker. He has a manual folding wheelchair as well as an electric wheelchair. By using his garage door opener, he can get into the garage when the bus drops him off. He requires oxygen for any change of altitude, thus airplane travel can create nightmares!

Andrea's favorite assist for Spencer is a 2008 wheelchair van.

Sergio Cardenas

Sergio has used a number of various braces, including a whole body brace to straighten his body and to help him to stand up, as well as a neck brace to keep his neck straight and hold his head up. He wasn't able to control his head until he was two years old. When he wore the whole body brace, Maria Elena had him sleep beside her bed, so she could rotate his body from to one side to the other during the night, because he would become uncomfortable. That brace was equipped with locks, and was meant to help stretch his body.

Sergio had braces to help to open his hands, and Afos (ankle-foot orthotics) that were supposed help him walk, but after two years, his doctor responded to his complaints that they were too heavy, and discontinued them to see if that would help his mobility. When he had a torso brace, which was meant to keep him straight to help him with his back and neck pain, he couldn't breathe well, or even move, sit, or walk. His doctor discontinued that also.

In 2004, Sergio started Pre-Kindergarten. He walked while at school and used a stroller for long distances. Then one day, he simply stopped walking and nobody knew why. They assumed he was tired or it was too painful. The school staff started complaining about the stroller, so Maria Elena took him to his doctor to try to figure out if he was just a boy who wanted his mother carry him, or if his disease was what made him stop walking. When the doctor saw his condition, he prescribed a wheelchair for Sergio's use. It took them a long time to comply with all of the insurance requirements for qualification. During that time he was able to use a borrowed wheelchair. In 2005, when Sergio was five years old, he received his own wheelchair. He likes to walk, run, and dance whenever he possibly can, but with the passage of time, that has become more difficult. He doesn't like to use the wheelchair when he is inside his home but he uses it everywhere else.

Sergio needs glasses to correct his vision and hearing aids to compensate for the loss of hearing in both of his ears. In addition, he has obstructive and central apnea, and has a CPAP (Continuous Positive Airway Pressure) machine to help him with his breathing problems.

Zachie Haggett

At the age of five, Zachie's breathing became much louder and heavier than it had been previously. He had always been a loud sleeper and snorer but never usually during the day, until that time. The school staff also noticed his breathing, so he was seen by a pulmonologist. Initial testing indicated signs of labored breathing during the normal periods of the day when he was not really over exerting himself. However, Zachie was only five, and not being a good steady blower, he did not get enough air into the machine so it could not present a clear picture of what was going on.

In 2007, when Zachie had pneumonia, he used a nebulizer four times a day. His parents have found that it is best to have him use it immediately at the first sign of congestion, as it is the best defense against repeated pneumonias.

By the time Zachie was eight years old, and very small for his age, his parents decided that they needed to make physical accommodations for him to help him achieve greater independence. When they told him that they planned to have a tiny sink and a tiny toilet installed in his bedroom, he was so excited that he wanted to stay home from school to watch the progress. The plumbers were taken aback as to how low they actually had to hang the sink, but Brenda brought in Zachie to show them how small he really was! Then they understood. While it might seem like a small thing to do, Zachie's parents believed it to be a huge thing for him and his self-confidence. Once Zachie had a complete bathroom of his own in his room, he started washing his hands eight or nine times a day, and touting how old he was now that he could wash his own hands and go potty on a "big boy potty."

In 2008, Zachie received new wheels for his wheelchair so he could propel himself. He exhausted himself by doing just that; he was so proud. His body gets stiff, no matter how much walking he does or doesn't do - he is just stiff all the time.

Also in 2008, Zachie was using a walker after surgery, taking his first independent steps, and taking some steps away from the walker and on his own. He likes to be on his own two feet as often as possible. However, since he had a second hip fracture in 2010, Zachie is no longer able to walk and is permanently using a wheelchair. He has also become too weak to propel his wheelchair, making him completely dependent upon others. It has been heartbreaking for Brenda and John who see this turn of events as yet another emotional loss for their family.

CHAPTER SIX

PAIN – PERSISTENT AND PERNICIOUS

Everyone knows what it is like to experience pain. Fortunately, most of us don't have to live with pain on a daily basis, as those with many rare diseases do. ML affects the body in different ways with some of the pain being located in bones, some in joints, some muscles and in multiple organs. Various treatments are employed to try to alleviate the pain – some successfully, some not. All of those affected usually see many doctors when seeking some relief from the various types of debilitating pain they suffer.

Hayden and Sarah Noble

Both Hayden and Sarah Noble had constant chronic pain for many years, and although numerous different kinds of drugs were tried, nothing ever really worked very well. They were unable to find a treatment for the chronic pain that was caused by bone deterioration.

Hayden required the use of a wheelchair by the age of fifteen. Sarah was still walking at age twelve, but she was becoming very unsteady on her feet due to huge changes in her hip joints. While she had always had pain, the level of pain by this age was increasing to a point where walking was very difficult. Consequently she lost all mobility and was also confined to a wheelchair. Jenny and Paul wondered if she was heading down the same path as her brother with spinal cord problems, but in time they were to learn that once the children started puberty, their bone destruction increased rapidly.

In May 2000, Sarah was seen in Sydney, Australia for a complete medical work up to try to determine the reason she was unable to walk. They didn't know if it was due to spinal cord damage or bone disease. The testing results showed that Sarah had gross destructive bone disease in her hips, pelvis and cervical spine. Her left hip had completely eroded away, and her right hip was heading the same way. She was also in danger of experiencing fractures. Her body had reabsorbed a disc in her cervical spine, indicating severe Osteoporosis. Subsequently Hayden was tested in New Zealand where they found he also had the same kind of bone destruction.

While in Sydney they learned about a bisphosphonate drug called Pamidronate that was being used successfully in treating patients with

Osteogenesis Imperfecta (Brittle Bone Disease). It had never been tried in MPS or ML children before, so there were no promises of a positive outcome. However, due to the level and pain and bone disease in Sarah and Hayden, the family decided to commence treatment on a trial basis, believing that there was nothing to lose and everything to gain by letting the children try it.

In August 2000, they began treatment with Pamidronate being infused on a monthly basis. Within two weeks after the first infusion, the Nobles were calling Pamidronate their wee miracle. Both youngsters were totally pain free. Three months after the first treatment, Sarah was able to get out of her wheelchair and walk with a walking frame. Within four months she was up on crutches, and at the five month mark all other drugs for pain relief were discontinued. Both children became drug free, and their bone density had increased.

Before they started receiving Pamidronate, life had been fairly stressful for everyone in the family. They hated putting shoes and socks on Hayden, as he always cried out in pain. Showering Sarah was terrible, as every movement caused pain. They were unable to travel long distances, so family holidays were not an option. After completing twelve months of treatment the pain issues were resolved, and Hayden and Sarah could do much more for themselves than they could previously. Their parents were relieved and pleased to hear laughter instead of tears. They came to an appreciation of just how miserable their children had been previously, due to constant chronic pain.

In November 2001, Sarah stood up and took six steps unaided. Her parents never thought they would see that happen again. She regained her balance and could walk around the house without support. Paul and Jenny really enjoyed seeing the improvements in both Hayden and Sarah, but there was one change that shocked them the most. Two days after arriving home from the Australian Conference, where Jenny presented a report on the first 18 months of treatment, Hayden stated that he could stand up. Jenny looked at him and said that it was not possible, as he is a paraplegic, but he was very adamant that he could stand, so they got Sarah's walking frame for him. Much to his parents' surprise, Hayden stood with the walking frame and then proceeded to take four steps. Jenny can't describe the emotions that shot through her and Paul that day. She felt totally numb; it wasn't until the next day that the excitement and hope kicked in. But in the back of her mind was the question as to how this could be happening… paraplegics don't walk, let alone stand up. She wondered just what else Pamidronate was doing for the disorder. They noticed that Hayden had not had chest infections since commencing Pamidronate and his quality of life had improved dramatically.

Twenty months after Hayden and Sarah started treatment with Pamidronate, the trial study was completed. The data on the "Significant Breakthrough in the Management of Secondary Metabolic Bone Disease in ML" was presented at the Australian MPS conference in April 2002, the 7th International Symposium on MPS and Related Diseases, and 3rd Scientific Lysosomal Storage Disorders Congress in Paris in June 2002.

Hayden and Sarah's journey with Pamidronate became a major medical breakthrough in the management of bone disease in ML, and one that has given them both a beneficial quality of life. They remained on Pamidronate for seven years, during which time they experimented with the dose levels and frequency of administration. The results from those trials were mixed. After seven years of treatment they started thinking about where they should go from there. Always aware that they could not stay on Pamidronate forever, they faced two options: move onto a new age bisphosphonate, or start an oral therapy. Fortunately they didn't need to make that decision right away as Hayden was starting to exhibit spinal complications again and both young people needed to have some dental work done. They had learned that there were some limits to the amount of the drug they could take, and that it was necessary to stop treatments months before undergoing any bone surgery, as it would inhibit normal recovery. The Nobles are not having infusions currently, due to both of them having spinal surgery.

After Jenny's presentation in Paris to the International Lysosomal meeting in 2002, they received many e-mails from ML families who had heard about what they were trying to do to combat pain. In the space of twelve months they saw well over sixty patients worldwide using Pamidronate. It was wonderful for them to hear of the improvements in many of those children as well.

In addition to the bone pain, Hayden and Sarah have suffered from muscle spasms, headaches caused by instability in their cervical spines, and pain caused by Carpel Tunnel Syndrome.

Sarah has progressed from conventional pain management to looking at more effective drugs and is receiving care pain management by the palliative care pain management team. Pamidronate infusions might be used again in the future if they believe it will be beneficial, and if surgery is not needed.

I first learned about Pamidronate when Jenny Noble wrote about it for the MPS Courage Magazine in 2003. She referenced the Lysosomal Diseases New Zealand web site (ldnz.org.nz), where I read the entire story. I was fascinated. I knew that her children had more severe bone issues than Kelley, but I also knew that Autumn Tobey was having serious bone issues, so I immediately contacted her mother, Pam.

Autumn Tobey

Autumn was twenty six years old when she first learned about Pamidronate. Since she had already experienced some serious bone problems, she gave her doctor the information she had received from Jenny and asked to be scheduled for a bone density test. The results from that test indicated that her bone density was indeed below normal range, so they made the decision to start infusions of Aredia (another form of Pamidronate) in 2003. Autumn was glad that she had been told what she could expect after the first infusion, because she felt really rough for a few days, with aching bones, loss of appetite, vomiting and diarrhea. While initially feeling extremely tired, she felt much better after a few days. Her doctor decided that she would only need the treatments every three months and that has appeared to be sufficient. She received some relief from the pain with each infusion and her bone mass was increased. Although Autumn did really well on the infusions, she had to stop them when she started having surgery on her shoulders. Each time she had had surgery, it has been necessary to stop the infusions from three months before until three months after the surgery so her recovery wouldn't be compromised.

Patients with ML III have to have many infusions for many different reasons, and their skin becomes thick while their veins tend to become worn out. The nurses describe Autumn as being a 'hard stick' meaning that it is very difficult to find a vein that will work for the full time of the infusion. The more often the veins are used, the more often they shut down. They considered implanting a port so they would be able to access her veins more easily, thus making the infusions less troublesome for Autumn. When discovering that a port can also present complications, like infections that would likely go to the shoulder replacements, she decided against the implant. The nurses apply a numbing agent to her hands and they stick her there, where her veins are better than in her arms.

The treatment has been helpful in reducing some bone pain, but most people find that joint pain is not relieved, and Autumn has had a great deal of severe pain in her joints. Pain medications go just so far in helping to reduce that pain. Of course, strong medications are needed right after surgery. Autumn has learned the hard way that she has a bad reaction to some of them. The doctors try to alleviate her pain during the time of intensive physical therapy, which is extremely painful and difficult after shoulder surgery.

Autumn has suffered muscle spasms after the surgeon has taken down the big muscle to dislocate the shoulder joint and then reattach it, and it has to be protected until it can grow back. That usually takes about twelve weeks. And, as many others with a similar condition know, weather changes can make Autumn's whole body ache, and there is little that can be done to make her comfortable at those times. Pain control remains a constant issue for Autumn.

Allison Dennis

Alli also had the benefits of some pain relief after she started to receive Pamidronate infusions when she was seventeen years old. Initially she had wonderful results from the drug, both with the reduction of pain, and with improved bone density. Unfortunately this did not continue beyond three years, but she continued using Pamidronate for five years hoping that it would continue to help her. Regrettably, when a stress fracture in her right hip didn't heal well, it was necessary to discontinue the Pamidronate for a couple of years to let the hip heal properly. The doctors had discovered that the drug was making her bones too hard. Two years after discontinuing the infusions, she started taking an oral bisphosphonate but found that was not very helpful, so she decided to stop that. Unfortunately, her pain then returned more often. While Pamidronate helps to reduce pain in some patients, it also inhibits bone healing, so the patients end up with a catch–22 situation, causing a great deal of frustration. Alli did try Pamidronate again a number of years later for a period of six months, but found it to be of no real benefit. Clearly, some of Alli's pain was caused by other issues, such as the pain she feels in her tailbone. The surgeon has told her that he doesn't want to remove her coccyx, as those operations are rarely successful. They did try using nerve blockers injected into that area, but Alli didn't feel much relief from that treatment.

Pain has definitely become a major issue for Alli, being cruel and hard to manage. It has been a struggle to find doctors who have been prepared to listen and try to help. For quite a while the pain was disabling her more and more, and Alli was crying all of the time saying she wasn't making up the pain up; it really was there. Some ill-informed doctors, who would not listen, made her feel that there was nothing wrong and it was all in her imagination. During this time Alli lost a lot of weight and became more and more socially isolated, since she didn't want her friends to see her in such pain. Alli has always been very stubborn and very independent, pushing herself hard all the time. Trish decided to take her to a counselor to help her through that difficult time. They also searched until they found a wonderful pain control doctor who is also a Neurosurgeon and understands Alli's pain. She is now on many medications in high doses as well as many natural supplements. Hydrotherapy has also provided her with good results at times reducing her pain to a level that she finds tolerable. Still, there are times when she suffers muscle spasms and the pain spirals. At those times, she tries to distract herself by praying, watching movies and listening to music. When she is in pain she tries to focus on positives and pray for other people. And Alli always manages to smile - no matter what is happening.

Huddy and Sammy Anthony

Liz Anthony doesn't think that either of her boys have had sharp pain, but they have had to deal with chronic low-level pain. In his last five years, when he was mostly immobile, Sammy had pain that he tried to overlook. When Liz asked him if his hands ever felt numb or fuzzy he said sometimes they did. Also, he would have heart-racing sometimes that he would tell her about when it happened because it scared him. When he took a nap during the day, Liz wondered if he might have been depressed. What Sammy really enjoyed was football and baseball season, and his laptop! He enjoyed visitors and holidays. He had serious spinal scoliosis, like all ML kids, and his attempt to alleviate the pain caused by it might be part of the reason he became immobile.

Huddy is less able to express what bothers him than Sammy and instead becomes 'ornery' when something is bothering him. Liz asks him about pain and tries to assess him based on what activities he can no longer do. It is apparent that his back bothers him when he is sitting in a small chair for too long. Sammy and Huddy never did try Pamidronate infusions, as Liz and Tom never felt that the boys would gain any benefit from that drug.

Callie Nagle

When Debbie learned about Pamidronate she shared the information with Callie's doctors. One doctor thought Callie wasn't a candidate, but the orthopedic doctor thought she was. They decided to discuss it with a doctor who was prescribing Pamidronate for patients who had Osteogenesis Imperfecta (Brittle Bone Disease). One of the doctor's mentioned that, because it was still considered a new drug, the dosage hadn't been clearly determined and dosages had changed drastically over the years. Debbie felt the information regarding the drug was overwhelming, and they needed to take their time in making a decision. She questioned if trying Pamidronate infusions would drastically help with Callie's hips, thus being able to delay surgery, or if it would be more helpful to have the surgery first, and then the Pamidronate infusions.

As they explored the options, one doctor was not supportive of putting Callie on Pamidronate, partly because she was doing so well, and partly because they claimed that the long term benefits versus complications were not yet known. Since the Pamidronate settles directly into the bone, there was a fear that later in life one could develop auto-immune problems. Of course, that was extremely alarming to Debbie and Richard. It was different information than they had received from the papers written about using the drug with ML patients. They felt they were caught in a quagmire, leaving Debbie absolutely overwhelmed, confused and frustrated. She might have had a different outlook if Callie's condition was more severe at that time. They finally decided that Callie should have a bone density test followed by a meeting with a surgeon to discuss her hips.

Callie's bone density test results showed that she did indeed have bone loss, so that changed everything in their minds about what awful possibilities may be in her future. They planned to see the specialist who treats patients with Osteogenesis Imperfecta with Pamidronate and bombard her with questions as they ventured into unknown territory. By that time, fourteen-year-old Callie was experiencing extreme pain, which frightened her, so they started to think that there wasn't any alternative, since the future seemed very scary and uncertain when looking at it without any treatment.

The specialist ordered lab work for Callie, the results of which showed that her Vitamin D and Calcium levels were both low, so she ordered supplements of those, because she was not 'sold' on infusing Pamidronate for anyone who isn't having fractures. She promised to repeat the bone density testing after Callie was on the supplements for three months.

It was almost two years from the time that the Nagles started exploring the idea of Callie taking Pamidronate until the time of her first Pamidronate infusion. The doctor had gone back and forth with discussions about the amount to give and looking at the results of Callie's testing. She also said that she was not comfortable giving the full dosage as recommended, and she would give it every six weeks rather than every four weeks.

Callie's first infusion went well and she continued to have Pamidronate infusions every six weeks for about a year. It made a huge difference in decreasing her pain and overall stiffness. She told her mother that when she went to move, she no longer felt like an old lady trying to get up and walk! Her bone density definitely increased, so they were very pleased with the results of the treatment.

In 2007, Callie saw the dentist for a routine check-up and cleaning. It was then that her X-rays revealed that her molars had never surfaced and she had an extra tooth (where it definitely didn't belong) and there was also an issue with her wisdom teeth. They saw an oral surgeon to discuss those seven teeth and the possibility of their removal. He had researched ML prior to their visit and was very prepared and informed. He told them that he felt that it was just too risky to pull them because Callie had been receiving Pamidronate, which meant that her jaw might not heal properly. There was also a very high risk of the jaw becoming infected. That surgeon had approximately five elderly clients on Pamidronate who had experienced such a jaw infections. Debbie learned that there is a hospital in Boston that has assembled a team to deal more effectively with Pamidronate and oral surgery. Meanwhile, no one wants Callie to take the risk of having any surgery done on her jaw. Pamidronate has proved to be a very tricky drug.

Callie also has significant knee pain, due to an excessive build up under her knee caps, which causes them to "float" and Callie's knee's constantly snap and pop (out of place), causing her constant pain. She experiences a lot of pain in her ankles, and she has a bunion on her toe which has been extremely painful as well. Sometimes she alternates ice and heat which gives her some relief and she uses anti-inflammatory medications. There is no one treatment that will allow her to be pain-free.

Jennifer Klein

Jenny is one patient who has received significant pain relief from Pamidronate infusions. In 2004, she and her mother met with the doctors and presented them with the results of all of her tests. Next they met with the dieticians, and everything was a go. Then it was time for Jennifer's six-month appointment with her orthopedic doctor who had not been in the loop, because he flatly turned Jenny down in use of Pamidronate the previous year. At that point Terri knew that he had to be included, since he was the Senior Head Pediatric Orthopedic Surgeon at the hospital where Jenny was seen.

Jennifer's tail bone area had started to cause some problems, and he began discussing surgery. Terri felt that she just could not keep anything in and therefore prayed to the Lord, thanking Him that He gave her the time, initiative, and support from the ISMRD forum, and the knowledge, needed to challenge the doctor. She pulled out all of her information and packets from her folders and proceeded to spread them out on the counters among two doctors and five medical students. By the time she was done, ten minutes later, she had every one of them convinced that the only appointment she wanted for Jenny was with the best Pediatric Endocrinologist they had. She wanted a doctor that was willing to look 'in the box' and then perform 'outside of the box!'

The doctor was shocked to see Jennifer's bone scan results, so by the time Terri was done she had an urgent referral to the doctor who was the first at that facility to use IV infusion Pamidronate for children with bone cancer and also one other Mucolipidosis patient. He said that he was familiar with the procedure, since the girl had been referred to him for possible surgery. Terri was shocked and asked if he was telling her that there was another patient there being treated with IV Pamidronate. He affirmed the fact that she stayed overnight for her infusions.

This left Terri feeling angry because she had been trying to get that doctor to listen to her for a long time. Now he said that they had just reported her good results recently! He added that he felt there still was not enough protocol, but did admit that Jenny's bone scans said it all. Terri told the doctor that either he was with them or not on their team; she wanted to think he would want the little ballerina to still be a special patient of his. Finally, the date was set for Jenny to receive her first Pamidronate infusion. Due to patient confidentiality, they were unable to learn any more about the other ML patient being treated there, except that she had ML II.

Jenny had infusions for many months, receiving very positive results both in an increase of bone density and a decrease of pain. However, that treatment was suspended after she broke her hip twice. About eight months after her hip healed she started on a different regime, taking a lower dose, and on a quarterly, rather than monthly, basis. Jenny sees both an orthopedic doctor and an endocrinologist. Her hip socket has been destroyed by the disease, so that a replacement is not a possible solution. Terri believes that Jenny receives a secondary benefit from the drug in that it has made her stronger and more able to fight off respiratory infections. She has had less breathing issues, colds and flus. But she is concerned about the endpoint and the effect the drug might have in ways other than the increase of bone density and pain relief. Jenny wants to avoid using narcotic drugs, so she uses analgesics for the pain that is not relieved by Pamidronate. Therefore, she must live with pain that both precedes and follows every operation and every hip break while the disease causes degeneration in all of her joints.

Andre Andrews

Upon learning about others with ML using Pamidronate infusions, Jane discussed the idea with Andre's geneticist. After looking into it, the doctor didn't think that it would be a good idea, due to the possible side effects.

Although Andre does have intermittent pain in his knees, he is seen by orthopedic doctors, and uses Lidoderm patches for some pain relief. Jane sees her son as a real trooper who accepts what he has been dealt and doesn't complain.

Joey Nagy

When the Nagys decided to investigate the idea of starting Joey on Pamidronate, his genetic doctor was very cooperative. She ordered a bone density test, and said that if that showed loss, she would have no problem starting him on the drug. Linda sent her all the information she had received from Jenny Noble. On the other hand, Joey's orthopedic doctor appeared to be on the defensive when Linda called to talk to him about it. He said that he had never heard of that drug being used for ML and he couldn't understand how it could help at all. Linda was surprised because they were actually using Pamidronate for their children with Osteogenesis Imperfecta (Brittle Bone Disease). It appeared that he did not like her suggesting what drug to try. He said he would like to see some supportive literature and asked if Joey was having fractures every day. Linda questioned if it might be able to help his hip joints, since his sockets were poorly formed. He said, "No," and continued to be negative.

Joey's first bone scan results showed that his bones were basically normal except for osteopenia in the left femoral neck. They contributed this osteopenia to the hip surgery he had when he was three. Linda disagreed, saying that he never had a socket to begin with. She received a brush off, with them declaring it was wonderful news, and they didn't have to worry for a while; they could repeat the scan in three years. Linda was thinking that it might be time to find a new doctor; at the same time Frank's company was offering enrollment choices of insurance coverage. That was fortuitous, as they were able to choose a different insurance company, and transfer to the hospital they believed would be more beneficial for Joey's care. Linda was able to talk to a genetic doctor at that facility that specialized in Metabolic Diseases. She had read all of the papers on Pamidronate and ML, and was well aware of the protocols to be followed.

Joey's first appointment with the new doctor went well, and tests were ordered. They took a blood test to check the level of osteoclast activity in him. This is the activity that destroys the bones. They also took full skeletal x-rays and a bone age. These were the tests that his doctors originally refused, due to insurance issues. After these tests and a new bone density test were completed, the doctor decided how much Pamidronate to infuse, although his decision was not taken lightly. Linda was very happy to finally have an agreeable doctor, after their previous battle. They believed that Pamidronate was the best way to prevent more damage to his bones as he entered into puberty.

All went well with Joey's infusion despite the fact they had to stay a little bit longer in the hospital than expected. They didn't have any problems getting a line in, although he hates that part. Saturday morning he woke up achy all over and was moaning that he felt weird. Linda insisted on using alternating medications, as she had been advised by others who were receiving the treatment, and that worked. They drew the blood at the required times, with Joey being scheduled to go home on Saturday night after the last draw, but that was when his calcium level dipped below normal. They repeated it a half hour later, and it dropped again. So he had to spend another night while he received some calcium in the IV. The following morning he was back to normal and they left at noon.

The following week was a bad one for Joey. Monday evening, he had to be carried to bed, because his legs and feet ached so much. When he tried to return to school with a medication that would last for eight to twelve hours, that didn't work; he returned home crying. Joey is not given to complaining, so Linda knew it was serious. The school had agreed to administer the medications to him on the prescribed schedule after the doctor faxed an approval to them. It hurt Linda to see him suffering like that, but she kept reminding herself of the end result.

After receiving Pamidronate a total of ten months, Joey's results showed he had a 24% increase in the bone density in his arm and a 35% increase in his spine, taking him well above normal bone density. However, Pamidronate did not turn out to be as good for Joey as Linda had hoped it would be. Prior to starting infusion, he was pain free, very mobile, he could run very fast. After his first infusion, things started to change for Joey, and he was becoming a different child, being in constant pain all of the time. He was even struggling in school while dealing with the pain and had to go to the nurse's office every day to rest and receive pain medicine, for which he eventually required narcotics. He also gained about 40 pounds during the 10 months he was receiving Pamidronate infusions. Linda complained almost immediately to the administrating doctor, saying she felt that Joey was a different child once Pamidronate was started. Having no experience with this, because all of the feedback she knew of was positive, the doctor didn't believe it was caused from the Pamidronate. Linda consulted with a number of other doctors. Some agreed with the administrating doctor, but one doctor did tell her to discontinue use, since it was being proven that Pamidronate was not beneficial for every child. After it was discontinued, Joey's pain eventually lessened a great deal, but never completely disappeared. Linda believed that the added 40 pounds of weight put more stress on his joints and limited his ability to move and exercise, so it created a vicious cycle.

Eventually, in high school, Joey was able to lose the extra weight and his mobility improved. Linda firmly believes in keeping him as active as possible, making him walk whenever possible, which is against the advice she has received from some others. However, she found that when she followed others advice, it only made him stiffer, so she will continue to do what she feels is the best way to help Joey keep moving to keep his weight under control.

Anna James

Brent and Jackie thought they should explore the possibility that nine-year-old Anna could obtain some relief from the pain she was experiencing when they learned about Pamidronate. They were excited to find a wonderful Endocrinologist at a nearby hospital who was very familiar with Pamidronate, as she used it to treat several children with Osteogenesis Imperfecta (Brittle Bone Disease). She read the reports about the positive results that some ML patients had experienced, and thought it would be a good idea for Anna to try it. Bret asked her if there were any drawbacks and her reply was that there is only one case of a child with Osteogenesis Imperfecta (in Montreal) that was given such high doses of Pamidronate that as he grew into adulthood, his bones would constantly fracture. She said that it is a very small risk in children with ML, especially since the dosage was much lower. She ordered a dexascan for Anna, which showed that she, indeed, had lost some bone density. She also remarked about the fact that Anna's hip bones looked quite rugged as opposed to smooth. That didn't come as a surprise to Anna's parents at that point, since they had learned that was a common symptom with ML children.

Before they were able to make the arrangements to start the Pamidronate infusions, the doctor ordered Miacalcin nasal spray for Anna to see if it would help control her pain in the interim.

Anna was able to receive Pamidronate infusions every month for about three years. Her Endocrinologist checked up on Anna constantly. It appeared that Anna did receive some pain relief, but after three years the doctor felt the treatment should be discontinued, as she became concerned that it would make Anna's bones too brittle. Jackie thinks that she might have been spared some of the pain that many others had during their teens, since she had her treatments before she reached puberty.

Four years after her last Pamidronate infusion, Anna started to suffer quite a bit of pain in her shoulder and neck. The doctor has not been able to determine what is causing the pain, since nothing has shown up on x-rays and MRIs other than some inflammation in her shoulder. She takes a medication every day to try and reduce the inflammation and bring relief from the pain. As Anna has been getting older, she does seem to have more bone issues. Her mother says that she deals with it quite well… perhaps better than she would in the same situation!

Spencer Gates

When they heard about ML patients being helped by Pamidronate infusions, Kevin and Andrea brought that information to Spencer's doctor, and asked for bone density tests. Andrea was anxious to have Spencer approved for the drug because it had been so helpful in relieving the pain of others.

After the results of Spencer's bone density test revealed that he had severely decreased bone mineral density, his doctor referred him to an Endocrinologist. Andrea asked if they could possibly see a doctor who she knew to be a Lysosomal storage disease expert, and was pleasantly surprised to learn that that he was available. He told them that he was very familiar with Pamidronate, but he thought they should try an oral bisphosphonate first. Andrea said that she was under the impression that IV infusion was the best treatment, but the doctor said that he wanted Spencer to try the oral medication once a week first. She mentioned that she had concerns about stomach side effects of the oral medication, and he said that he could take a dose of an acid-reflux medication the night before to abate the heartburn. They agreed that in four to six months, if the oral medication was not working, they would start the IV of Pamidronate.

Spencer seemed to tolerate the oral medication with very few side effects. After six months, there was an increase in his bone mineral in two areas. But it was very disappointing to them that he still experienced pain daily. The decision was finally made to start Spencer on Pamidronate. The first infusion went well, and they waited for Spencer to feel relief from some of the pain, but they were disappointed. The same thing happened the next month. By the time he completed his fourth infusion they still had not seen any of the benefits they had hoped for. His pain was the same, if not worse, on some days. After each infusion Spencer asked if this time he would feel better. They decided to increase the dosage for his fifth infusion. A few nights later, before he went to bed, he said the pain in his legs was at a nine, on a scale of one to ten. His mother was heartbroken to see he was still so uncomfortable. She felt frustrated and started to wonder why they even began that treatment. After every infusion Spencer asked again if this would be the one that would make his pain less. They went ahead with monthly infusions, with increases being made in the amount of the drug. While his bone density did increase, his pain level did not decrease, as it had for some other ML patients. Clearly, it was time to explore other ways to help Spencer handle the pain. Andrea was very disappointed that Pamidronate was not the wonder drug for Spencer that it had been for some others, but she did wonder if his pain would be worse if he had never used the drug. There is no way to know for sure.

In 2007, Spencer was on a third medicine for his constant pain. He told his mother that he just wanted one day without pain to see what it would be like to be normal. His pain management doctor tried nerve blockers which provided some relief, since he hates to take narcotics because there are so many nasty side effects. Spencer has been on pain medications since the third grade. Actually he spent all of that year carrying a 'special' red cup with him, because he was getting sick almost every day. That was the year they tried every kind/type/brand of medicine to try to reduce the pain, keep him awake, and not make him throw up! It truly was a terrible time. It took many years, doctors, and medications until they finally found a way for him to function with the pain being reduced, even though it has not been completely eliminated. To them, the entire process felt like it took forever!

Sergio Cardenas

Sergio was four years old when he started complaining about pain, but it was difficult for Maria Elena and Gustavo to realize what was happening. They heard him crying and hitting himself against the wall of his room in the middle of the night. At that time Sergio could barely talk. He had a limited vocabulary and didn't even know the word pain. Maria Elena was alarmed when she went into his room, turned the lights on, and saw the blood on the wall. After there were a few episodes like that she felt it was not fair for Sergio, or his brother and sister, for that matter, to experience those nights. She decided to stay in his room while he slept in order to discover what was happening. She stayed awake all night with a small light to watch him and all of his moves. She was able to determine that the boy was suffering from pain. Thus, she embarked upon a project to teach Sergio the word pain and its meaning. She drew a big boy on a cardboard where she illustrated the human body parts and organs to teach him where they were, and the names of the parts so he could tell her where his pain was. It took about three or four years for him to learn everything necessary for him to express how he felt using words. It is still hard for him to let others know how he feels, since he thinks the pain is part of his body and his life. It wasn't until he was about ten years old that he realized others don't feel pain. That makes his mother feel very sad.

In searching for ways to help alleviate Sergio's pain, his doctor believed it would be worth it for him to try using Pamidronate, since it has helped others with ML. At first, it appeared that he was receiving some benefit from the infusions, but then the pain started to become worse after each treatment. The doctor tried higher doses and more frequent infusions, all to no avail. It became apparent that Sergio was not going to benefit from the drug in the way that some others had. The doctor and Maria Elena agreed that Sergio appeared to be getting more mobility with his therapies than with the infusions, so the Pamidronate was discontinued.

Sergio suffers from bone pain, muscle pain, and joint pain, and he is able to identify the difference in each one. For this each type of pain, he needs to use a number of different types of medication, sometimes they help, but as it is with many other ML patients, very often the pain is too severe to be totally eliminated.

Zachie Haggett

Brenda was pleased to discover that there was a doctor in Syracuse using it for children with Osteogenesis Imperfecta (Brittle Bone Disease) after learning that Pamidronate was being used to help alleviate pain in some ML children. She and John considered the decision for Zachie to have the treatments to be a no-brainer, since they only want to help his quality of life. He had a medic-port implanted for the infusions.

When they were shown Zachie's full skeletal x-rays that were taken before he began the Pamidronate infusions, Brenda and John saw with their own eyes how horribly the disease had already affected him. There were no visible signs of bones at all. The films looked like puffs of smoke. They were told that their son already had the body of a ninety-two-year-old. It was hard to understand how he could be that bad at such a young age, since most of those with ML III didn't have as many problems until they reached puberty. Later they were to discover that his correct diagnosis was shared with very few others. The condition is now termed ML II/III, since isn't quite as severe as it is with ML II patients, but much more severe than with ML III patients.

Zachie can be quite articulate when he is not happy. And he was not happy for those first few infusions, but he gradually got over it once he learned how much spoiling he got at the "hosipal" with nurses at his beck and call for an entire day. It was agonizing for his parents to watch the pain he went through in the beginning. When the infusions brought about some pain relief for Zachie, as well as an increase in bone density within a year, they were pleased.

In time it became a struggle to find the right dose of Pamidronate for Zachie. He received pain relief for about three weeks, but by the last week of every month, he became miserable before the next treatment. They attributed that to him being very uncomfortable, since half way through infusion day he turned back into Happy Zach right before their eyes. Hoping to keep his breakthrough pain at bay, they started giving him infusions every three weeks instead of four. They realized that it would not have any effect on the continuing storage aspect of his disease, but felt it to be far more important that he be comfortable.

In 2007 Zachie's infusions were discontinued due to extensive growth in one area of his hips over everywhere else in his body. That was discouraging, as the drug was what had relieved most of Zach's pain for the preceding years. The day of his last infusion of Pamidronate was a bittersweet one. Brenda had hoped when Zachie first began the drug that it would help him for years to come. Now they would have to find something to decrease the pain that would not do further damage to his already taxed liver and spleen. Do to the nature of Zachie's disease, the storage that had already built up in his organs made it a little more complicated to find the right pain medication.

On the day of Zachie's last infusion, he schmoozed his way through the staff. They were always wonderful to him, and on that day they showered him with gifts and a lot of extra love and his favorite... kisses from beautiful ladies!

By February of 2008, Zach was able to begin infusions of Pamidronate again. Within a few months, he was feeling much better, with much less pain in his legs and feet. As the years have progressed, Zachie has used Pamidronate as much as possible, in and on-again, off-again fashion, depending upon the effects it has had on his bones.

Kelley Crompton

We didn't become serious about Pamidronate for Kelley until we started to see that some patients were experiencing a decrease in pain as a result of the drug. Kelley had a bone scan in 2000, where her bone density was considered normal, but we asked for another scan after learning about Pamidronate a few years later, and we discovered that Kelley had indeed lost some bone density. I then commenced upon doing my own research. The only negative that I could find was that a child was given large amounts of the drug without proper follow-up testing, and he had some severe problems as a result.

But, I also learned that a randomized, double-blind controlled trial had been done with adults who have had hip replacements. This was done in an attempt to keep the integrity of the bone so the implant wouldn't fail, and the initial results were good enough to call for more long-term larger-scale clinical trials.

Since Kelley has had both hips replaced (and both knees, also), we felt it was important enough to discuss this with the doctors. The first one we spoke to passed it off, but we didn't stop there. It took a few tries to get someone to pay attention, but then, the doctor in charge of the pain clinic where Kelley was seen did pay attention, and in fact, became excited about the prospect. He said that they had been using Pamidronate infusions for years to treat cancer patients, and he didn't see why they couldn't also try it on Kelley both for the integrity of her bones, and to hopefully reduce the pain she had been experiencing, particularly in her temporomandibular joint, which was extremely painful for her. She had been told that surgery for that condition was too experimental for her to consider.

When we saw the endocrinologist she had been referred to, I took along all of the information I had received from Jenny Noble as well as Kelley's bone density test results. After he ordered blood work, and received the results, it didn't take him very long to say that he saw the value in having Kelley try it for six months, after which he would re-evaluate the situation. Although I don't think we would have taken this path if Kelley's bone density test had been normal, I was grateful to Jenny for alerting us to the issue.

A new bone scan in May 2007 showed that the density in Kelley's spine was back to within normal limits. Kelley did seem to receive some reduction in pain during the first six months on Pamidronate, and was able to cut down on pain medication. It was very difficult to judge after that, because she had many other health issues.

However, the amount of pain she experienced, due to the various parts of her body being attacked by the disease, progressed as the disease progressed, and she reluctantly reverted to the use of narcotics to try to reduce the pain. At first, she needed them only occasionally, but the day came when she needed continuous dosages of narcotic medications. Although she could put up with some of them making her feel sick in her stomach, she was crushed when she realized that she would never be able to stop them without suffering severe pain. Kelley wasn't one to complain. She didn't cry often, but she did shed tears over the fact that she would have to reluctantly give up driving a car.

It is not within the scope of this book to cover the topic of paying for the drugs being used to help alleviate the pain these patients experience. Insurance often covers the use of some drugs for certain reasons, but those with rare diseases usually do not fall within those guidelines, so all of the patients who have rare diseases often face daunting problems in trying to find relief from relentless debilitating pain, while their loved ones feel helpless when watching them suffer. Hopefully there will be a solution for this problem before too long.

CHAPTER SEVEN

SURGERY AND MORE SURGERY

It is common for children and adults with ML II and ML III to undergo surgery numerous times. While bone and joint problems are the most common ones, the disease attacks bodies in such a variety of ways that just about any type of surgery might be necessary. Anxiety is always high when an ML child needs surgery, since many children with these diseases actually die from unexpected complications that result from the procedure.

At the time this book goes to print, those whose stories are related within have collectively been to the OR at least 140 times. While some of those have been short visits, some have been very long and extensive. One of the remarks we have all heard from the surgeons, more than once, is, "I've never seen anything like it!"

It is not possible describe all of these operations, but a sample can be found here in addition to those that have been discussed in other chapters.

Kelley Crompton

Shortly after our daughter, Kelley, received the correct diagnosis of ML III, she was referred to an orthopedic specialist at the medical center to determine if she had Carpal Tunnel Syndrome since that condition has often been found in children with the similar condition of MPS. This is a condition in which the nerves that lead to the fingers are cut off by compression within the wrist area in what is commonly called a tunnel. This tunnel must be opened to release the nerves and remove any accumulation of unwanted material so the feeling to the fingers will not be lost. Many children with Carpal Tunnel problems can't determine hot from cold; they complain of sore hands, pins and needles, and aren't even aware that they don't have normal feeling in their fingertips.

We had no idea that Kelley had lost feeling in the tips of her fingers. She never complained about it, because she didn't know that it wasn't normal. She couldn't remember if she ever had feeling in them. Once we learned that, we understood why she often dropped things and sometimes had cuts in her fingers and bled profusely before being aware that she had been cut.

The surgeon said that he didn't expect that the feeling would return as a result of the surgery, but if she didn't have the surgery the situation would progress. He planned to do one hand and let that heal completely before doing the other.

Kelley had to have a test called an EMG before the surgery. During this test, needles are stuck in the ends of fingers to send electrical impulses up the fingers. The purpose of this was to determine how much feeling was still present and how much was missing. They sedated Kelley ahead of time. Nevertheless, it was a very nasty test for a child to endure. I had not been able to prepare her first, because I didn't know what it was like. I thought it would be painless, like an EKG or an EEG. I was disturbed that I was out in the hall, rather than with her, when she was going through it. Had I known that she was going to feel a shock every time they sent an electric impulse through each finger, I would have insisted on being in the room with her.

I was able to prepare her for the surgery. My own experience with surgery, plus my hospital experience helped me to let her know what to expect. I believe that people can cope better when they are prepared by being told the truth. When someone says to a child, "This won't hurt," before giving a shot, all they do is create distrust!

We left Kelley at the hospital on a Sunday night. This was before the time that parents were encouraged to stay with their children. The staff really did not want to have parents around. We were *tolerated* during visiting hours. The surgeon said he would call me at home when the operation was done, and I could head into the hospital while she was in recovery.

Since this was Kelley's first operation, our anxiety was high and the wait seemed very long. In retrospect, it was a relatively short time. It went well, and the surgeon said that he took "a lot of junk out of there."

A few months later, Kelley had the operation for her left hand. The surgery went well, and recovery from this operation also went very well. Then the surgeon wanted to do another EMG for a comparison with the one done before surgery. Kelley knew from that experience what she was in for and was very nervous on the way in to the hospital. They gave her a sedative, and I insisted on being in the room with her this time. I was told they didn't usually allow that, so I said that if they wanted to do the test, they were going to have to put up with me. While the test was taking place, I did what I could to make it less traumatic by talking to her as much as possible without interfering with the testing. I started by asking her where she wanted to stop to eat on the way home. She mentioned a restaurant at a particular mall with a waitress who seemed to delight in taking care of Kelley. I said that was a great idea, and we chatted about that waitress. Then I noticed that her purse had become rather worn, so I suggested we would shop at the mall to find her a new purse. We discussed that in between the electric shocks. Next, she brought up the idea of buying new shoes. I was relieved that the test did not last any longer. It turned out to be an expensive one for me, but not as upsetting for her as the first one had been. I felt that the doctor decided that having me there was not such a bad idea, after all. Kelley was very cooperative during the entire procedure.

Little did we know at that time that she was to enter the surgical suite for many different reasons in numerous hospitals more than fifty times as she went through life. Some of the surgery was to decrease pain and make her more comfortable, but some of the surgery was actually life-saving. When problems with her trachea severely restricted her ability to breathe, causing her to spend many days in Intensive Care following surgery, we found a very experienced surgeon who tried to repair her trachea. That turned out to be one of those, "Never seen anything like it!" times. He couldn't really repair it, but he did find a way to keep her trachea from collapsing, thus allowing her to breathe for many more years.

Autumn Tobey

Autumn also had many surgical procedures over the years. The first was a hernia repair. Then she had a Spinal Fusion of her lumbar spine when she was twelve years old, Carpal Tunnel release on both hands at thirteen, and a tendon release in her right hand two years later. She had a synovectomy on her left ankle when she was sixteen and an ankle fusion on her right foot a few years later, followed by an ankle fusion on her left foot shortly after that.

When Autumn was twenty-five, she was told that her left thigh bone was dying. It was discovered that the blood supply had stopped at a certain point in the bone. Therefore the bone from that point on was not nourished, and it simply crumbled. She needed to have an allograft, but it took time to schedule the surgery because it was difficult for them to find a match, due to her bones being so tiny. There was a further delay when they thought they had a match with what turned out to be bad bone, so the search was on again, and a replacement was found about three weeks later.

When the surgeon actually got into her leg, he found the damage to be much greater than he expected. When she was discharged, the doctor forgot to order antibiotics for her to take at home, so by the second day she spiked a high fever. That frightened Pam who stayed up all night working to bring Autumn's fever down, hoping that no damage had been done to the allograft. The next day she insisted on a prescription for an antibiotic. Autumn was housebound for a long time after that surgery. She spent eight months on crutches, and was admonished that she should not put one ounce of weight on that leg for the first three months. Pam was relieved that her own beauty shop business was right next door to her home, so she could be there for her daughter throughout that time. Six months after surgery, Autumn said that the fact she could put full weight on her leg with a brace and crutches was the best Christmas gift she could receive.

It was a great loss for Autumn when Pam's father died, since the two had absolutely worshipped each other. Seven weeks after his death, Autumn had surgery again, and when Pam went into the recovery room, Autumn looked at her, smiled and said, "Granddad never left me alone in the operating room; he was in there the whole time."

Pam believed that indeed he was, because her recovery from that surgery was the best she ever had.

Both of Autumn's shoulders have caused her a lot of trouble, and a great deal of pain. The growth plate on top of her shoulder was fragmented, so the surgeon attempted to repair it. At first he thought he would remove the damaged part, but the MRI showed that he would have to remove about a third of the bone, which would leave her shoulder totally useless. Therefore, he grafted bone and used screws and wires to repair it. Her shoulder was then immobilized for six weeks. After healing from that surgery, she had a total shoulder joint replacement on the same shoulder. The surgeon wanted to wait until she was older, but he said that it was so deteriorated it might not be repairable if he waited. The bone in the shoulder was so badly damaged that he couldn't use it for the fusion. He said that the bone didn't look like a bone, in that it was very fibrous. He didn't want to take bone from her hips, because they had already been harvested three times previously. He used an allograft for the fusion. It was also necessary for him to remove some of the muscle. Next they did a hemi replacement of the left shoulder, which meant that they replaced the ball with metal and resurfaced the socket. Six months later, they did the same to the right shoulder. The doctor had thought that the left one was the worst one, but he was shocked when he discovered that the right one was really severe. That situation belonged to the, "Never seen anything like it!" category. Only a year later, the pain became severe in Autumn's right shoulder again. An X-ray revealed that it was dislocated; the muscle had failed. That meant that there were only two options left for her - fuse it, or do a reverse shoulder joint replacement. She had barely enough bone left but it was enough for the reversal.

When the left shoulder surgery failed, Autumn needed a complete reverse replacement once again. They decided that while she was already sedated, they should perform another carpal tunnel release done on both hands.

She has also had a defibrillator implanted to control her irregular heartbeats and prevent sudden cardiac arrest.

To date Autumn has had fourteen major operations, and the reality is that more can be expected, while she is hoping that the shoulder repairs will not fail.

Allison Dennis

When she was very young, Alli experienced recurrent ear infections as well as daily nosebleeds, so she had her first surgery when she was four–years-old. They inserted grommets in her ears and cauterized her nose. Up until the age of twenty-two, Alli had only relatively minor surgeries.

In 2005, Trish took her to the doctor for a check-up. At that time she asked the doctor to check her knee, as she had been complaining it was sore. He sent her for X-ray's, after which they were called back into the doctor's room, where they were told that Allie had a stress fracture in her hip. She had had been walking during that time, using her wheelchair only for distances.

After seeing a bone specialist, she had to wait a further six weeks before the operation. Then, six weeks after the surgery, they discovered that it was not healing. Alli was receiving Pamidronate infusions at that time, and they realized the drug was making her bones too hard, hence the stress fracture. It was necessary to wait another four months after discontinuing Pamidronate infusions before enduring another hip surgery. This time the doctor performed an osteotomy to realign the hip socket, so that the ball and socket fit together properly, and inserted an adult ankle brace, as her hips were too small for adult size hip brace.

Alli hates to have surgery and has to work herself up to it. This is a difficult process, causing heartfelt emotions and many tears. To date she has had at least fifteen operations, many on her spine. To make matters worse, her scheduled surgery has been postponed four times, where Alli has prepared herself for surgery, and then told at the very last minute that it wasn't going ahead. When the surgery is cancelled, it feels to the family like everything comes crushing down with a huge thud, only to have to prepare Alli again when the next date is given. It becomes a very traumatic situation. For Trish, the hardest part is to see Alli go through the agonizing preparation for a surgery that doesn't take place, which leads to anger, frustration and almost defeat. These times are very difficult, not only for Alli, but for the whole family.

Alli's latest surgery was done to repair the lower screws in her spine that had become loose, causing massive inflammation, and a tremendous amount of pain. During the eight hour surgery, they injected surgical cement into the offending screws to hold them firm. It was a grueling experience for her, since it was all done under local anesthetic, which they had to repeat three times. Of course when they got in they realized it was a lot worse than they thought, and they had to do the other side as well. She had a very rough night with pain and heaviness in her spine and legs. She was in Intensive Care for five days afterward before being transferred to a regular room. The recovery from this surgery was a long one, with severe pain, due to the fact that it caused considerable inflammation and bleeding. Alli hopes and prays that she will not need any more surgery.

Hayden and Sarah Noble

As a small child, Hayden had numerous ear, nose, and throat infections, consequently having his tonsils and adenoids removed, and grommets inserted in his ears to prevent glue ear. His parents didn't attribute those problems to ML III, as they were told that many young children have such issues. About two years later he had his second set of grommets and to this day he still suffers with glue ear.

Hayden was eight years old when the family traveled to America and then on to England to attend the first International MPS conference, hoping to understand a little more about ML III. While in America they discovered that both children had Carpal Tunnel Syndrome, and they also received advice on the kind of spinal surgery that Hayden should have to stabilize the Kyphosis/Scoliosis between his lumbar and thoracic spine.

When they returned to New Zealand, the doctors were astounded at the severity shown in the results of testing for Carpal Tunnel Syndrome. Surgery was scheduled. This was the family's first experience of health professionals not understanding the condition.

Hayden's first Carpal Tunnel operation was performed in November 1990. He seemed to have good results, but in March of 1993 he was back for a second operation. At that time he also had a Bilateral Metacarpal Release, as he was starting to have clawing of the hands, and they were becoming useless. He experienced improvement, but by 1995, yet another operation was necessary. This time it was to be a little different, as it was thought that these children could have a compression either above or below the Carpal Tunnel. It was suggested that the operation be carried out further into the hand to look for compression. The Nobles had to convince the doctor that this is what he should do, as they did not want Hayden to go back for yet another operation. Compression was indeed found in the palm of his hand. The results of that operation were fantastic, and Hayden has not required any further surgery for Carpal Tunnel.

When Hayden was nine years old, Jenny traveled to Auckland with him for major surgery to stabilize the curvature of his thoracic spine, which was successful. Then, by 1996 Hayden started to have quite a lot of neck pain that caused headaches and chronic tiredness. An MRI scan showed that the cervical spinal cord was swollen and there was no spinal fluid protecting the cord, indicating cord compression for which a surgical release was necessary. Since it was a fairly major operation, Jenny and Paul were concerned that he was becoming more at risk with anesthetics due to his neck problem and thickening of the airways after having so many operations in such a short time.

The surgery was considered successful, and Hayden seemed to be progressing very well until about fifteen weeks post-op, when he started to experience problems with his balance. They were frightened when he had to rely on holding onto anything he could just to stay upright. After extensive testing, it was determined that the disease had infiltrated his spinal cord and he was experiencing cell death of the nervous system. They were shocked to learn that the cord was swollen again and the spinal fluid was no longer around the top of the cord. Steroids were used to help reduce the cord swelling, and Hayden was able to walk with a walking frame. The steroids were stopped after two weeks and Hayden seemed to be making good progress, but two weeks after stopping treatment he began to fall again and was having problems with his bowel and bladder.

Treatment was recommenced and Hayden found his mobility again. That situation continued for one year, when it was decided that the drugs were doing more harm than good and treatment was stopped. By that time he was relying more and more on his wheelchair, so it seemed the right decision to make, since the disease did infiltrate the spinal cord and caused cell death of the nervous system. Unfortunately for Hayden, he became a paraplegic with ML III and a very high risk for any surgery.

Sarah has followed her brother with some surgeries, but has had to deal with more pain than Hayden has, due to him going into a wheelchair sooner than they expected.

Sarah also had her adenoids removed, grommets inserted and Carpel Tunnel surgery. Like Hayden, compression was also found in the palm of her hand. Sarah's recovery from the operation was positive, and she has seen continual improvement in her nerve conduction studies.

After Hayden had his cervical spinal fusion surgery, he wore a neck and upper body brace for six months to give him added stability and allow for the healing process to take place. That operation took quite a lot out of Hayden, which meant it took longer for him to recover. However he did recover and for a little while he enjoyed really good health.

Sarah has had a new ball inserted into her left shoulder. The surgeon has relied on the muscles holding it all in place. She has a little more movement than she had before the surgery, with an added bonus of her shoulder being pain-free!

To date, Hayden has had twenty-one operations, and Sarah has undergone surgery eight times, some of which is covered elsewhere in this book.

Huddy and Sammy Anthony

Sammy had two sets of ear tubes that did wonders for his health. He had constant ear infections previously, requiring a list of eight or so different antibiotics in one year, one of which caused an allergic reaction. It had been recommended that he also have his tonsils and adenoids removed, but his parents ruled that out. They decided to avoided surgery at all costs because of the probability of his narrow airway swelling and closing up, as has happened with many of those who have ML.

Sammy also suffered from obstructive and central sleep apnea for many years.

Callie Nagle

Callie has had a total of eight operations to date. She had two inguinal hernias when she was very young, one surgery at eighteen months, and the next when she was just over two years old. Her first successful Carpal Tunnel release surgery took place when she was six years old and the second when she was seven. Both were done on an outpatient basis, and went smoothly, even though she never went under anesthesia without a fight.

When she was in the third and fourth grades, she had surgery on each hand. Those operations were quite a bit more involved. The surgeon they sought out for all four hand surgeries was a specialist whom they liked and respected. He not only performed tendon transfers, but also attempted to perform trigger releases on four of her fingers to straighten them. Sadly, the latter produced no success. These extremely painful surgeries required Callie to stay overnight in the hospital each time. When she woke up, she had tiny metal rods coming out of the tip of each finger. Years later, she still sometimes talks about how some of the kids in school made fun of her. The first of the hand surgeries was truly a horrible experience right from the start. Callie was as scared as scared can be. She had such a fear of needles; Debbie didn't know how they would get through it all. It was never easy getting Callie to take medication. It didn't matter what the flavor or texture, it wasn't happening without a battle - ever! So much for the little cocktail they thought she would drink to help calm her down. One of the doctor's told her the IV needle would feel like a mosquito bite. Meanwhile Debbie was thinking, "Oh no, don't tell her that!"

After the surgeon, two nurses, a few other people, Richard, and Debbie held down the screaming child, they got the IV in, and Callie angrily declared, "That was a bee sting, not a mosquito bite!" She was right. This was after they put numbing cream on her one hour prior to starting the IV. This process was pretty much repeated for all four of her hand surgeries.

That hospital let one parent accompany their child in the operating room. Callie wanted her mom, so Debbie donned the scrubs and walked next to Callie as they wheeled her to the OR on a stretcher. With daughter crying the whole way to the OR, mother did her best to stay calm... until the nurse put that black plastic mask over Callie's face while she was fighting it off. The sight of her finally sleeping, with tears running down her face, still brings Debbie to tears as she recalls this incident. She'll never forget that image. Callie, now an adult, says that she still hates the smell of rubber (from the mask) because of that experience. Her memory is that everyone, including her mom, looked like aliens, all talking in slow motion saying, "It's all right." Debbie felt like she was falling apart as she walked out of the operating room, thinking that she didn't want Richard to see her like that and wonder what the heck was going on. She knew that he would not be totally surprised by her emotional meltdown, but she thought that she wasn't a pretty site for him to see after his being left in the waiting room. She'll always remember with gratitude two wonderful nurses consoling her walking her out to the waiting room.

Once Callie was out of surgery, her parents were brought in to be with her. She had ended up going into surgery later than scheduled, so it was late and the day surgery staff was wrapping things up. Callie was the only patient left in the post-surgical unit, and even though she was vomiting and experiencing a lot of pain, her nurse made it very clear to them that she wanted to finish her shift! She started going over the post-surgical instructions while Callie was crying in pain and still getting sick.

When Debbie told her that she wasn't taking Callie home in that condition, the nurse appeared to be very annoyed. So she called the surgeon to tell him what was going on, and he admitted Callie. Richard went home to sleep. Callie and Debbie arrived in her room around eleven o'clock that night. They were both sleeping when the nurse came in sometime after they arrived. She woke Callie by taking her blood pressure and temperature, but the thing that stands out in Debbie's memory is that the nurse never spoke to either of them. Callie woke up startled upon seeing a stranger over her, so Debbie immediately jumped up to her bedside and explained what was going on and tried to soothe her. The next morning Callie had the dry heaves and was still in severe pain. The nurse never oriented Debbie to the layout of the floor or how to order breakfast for Callie. When the next shift nurse came on, she helped to get Callie off of the medicine that was making her sick and told Debbie how to order food for her. They were most anxious to leave that place after that surgery!

After returning home, Debbie called the Director of Nursing and complained about the inexcusable nursing care and experience they had, and told her about the horrible pain Callie was in and how sick the medication made her. She was relieved that the director promised to put together a pain team for Callie in preparation for the next surgery. This in itself was the biggest improvement they could have asked for. As difficult as the following surgery was when putting her under anesthesia, the overall experience was so much better. They were thankful that the terrible in-patient experience was never repeated.

The two hand surgeries that required the metal rods in each finger also required removal of the rods. The first time they were removed was under anesthesia, when Callie was eight years old. After the second hand surgery, the surgeon felt that Callie was old enough to have them taken out in the office. They all agreed that it would be best to avoid anesthesia, unless it was completely necessary, but the Nagles had no idea what to expect. The doctor took a little pair of surgical pliers and commenced to pull each metal rod out of each of her four fingers. The rod went down through both knuckles, so it was a good two inches long. Debbie felt nauseous and light headed while watching. Callie screamed, not only in fear, but also in pain! Thinking that the approach was barbaric, Debbie was weak in the knees when leaving the office. That was one appointment that she wished she had missed.

In October 2012, after suffering many months of extreme pain in her hip, Callie finally had a complete hip replacement. After seeing three surgeons at three different Boston hospitals, where none felt confident that there would be a successful outcome in performing this surgery due to the severity of her hip dysplasia, Callie was referred to a surgeon in Baltimore. The two surgeons who performed the operation said it was very challenging, but also successful. They used a piece of her femur to graph to her pelvis to create a socket. The first few days post-op were pretty rough, but after a change in pain medications, Callie started feeling better and she was able to walk a little on crutches with the assistance of a physical therapist. Her surgery took place on a Monday, and they flew home that following Saturday! Callie had obvious hip restrictions. She was still experiencing severe pain, having only been out of bed for the first time on Wednesday. They felt that the successful car transfers, wheelchair transfers, and plane ride, with her legs dangling from the seat, were nothing short of a miracle. Debbie feels that God carried them home in His hands. Three months post-op Callie's pain had subsided; she was driving again, going for physical therapy, and best of all, Callie was able to go back to school.

Jennifer Klein

Jennifer had Rosella when she was very young, and required surgery for ear tubes twice as well as adenoid removal. This was shortly before she had back surgery when she was nine years old. Her doctors talked about trouble with intubation, and assumed at that time that her airway was swollen because it had been accessed too often. Jennifer's scoliosis was so severe that the surgery took about nine hours to complete. They accessed her spine from both front and back, using cadaver bone and screws to make the necessary repairs. The pain she suffered during that time was so severe that when it was time to take her home from the hospital it took four hours to move her from the bed to the wheelchair. They sent her home with a very strong narcotic drug which made her very sick, and she suffered significant intense pain for a very long time.

Shortly before her twelfth birthday, Jenny was hit with an infection that, at first, appeared to the doctor to be viral. She saw the doctor three times over the course of ten days, as Terri was convinced that something was not right and the doctor was missing it. She felt like she had to fight to get the right antibiotic for Jenny. At first the antibiotic seemed to work, but it could not keep up with the infection. The doctor ordered X-rays. Jenny's lungs looked fine the first time. However, a few days later it was clear that she had a full-blown pneumonia in her left lung, and she was admitted to the hospital. When they studied her blood, they discovered that she did indeed have a bacterial infection. In fact, Jenny had a new and aggressive strain of streptococcus bacteria that was causing the pneumonia! She had a fever for forty-five days, and was in the hospital for over a month, during which time she had four operations to extract the infections from her lungs.

This particular infection caused globules to form like large grapes that the antibiotics couldn't penetrate through. Therefore, she had a needle extraction in her spine hoping to remove enough of the infection, but that wasn't sufficient. Next, they had to do laser surgery to break up the multitudes of globules and then insert a chest tube that remained in place for four days. Later, she had to have the chest tube reinserted after removal in another surgery, and then she had more globules removed. Terri and Walt felt like they were living a nightmare. They fully expected they were going to lose their daughter during that time, as she underwent treatment, being allergic to some of the drugs, which caused severe vomiting, and losing twenty pounds in the process. Jenny didn't really turn the corner until after she returned home. Her parents believed their many prayers were answered when she made a remarkable recovery, and after three long months was able to return to school.

That experience taught Terri that she would take Jenny to the hospital right away if something like this happened again, rather than try one medication after another through the doctor's office. Jennifer has had a number of respiratory infections since then, but they have been able to get treatment for her in time to avoid a repeat of that frightening event. Jenny does have residual pain on the left side due to scar tissue left from the assault on her lungs during that terrible time.

Andre Andrews

Andre has had surgery to place tubes in his ears on more occasions than Jane can count – almost annually. Thankfully, this is the one procedure that is not a problem. It takes about fifteen minutes and doesn't require the kind of anesthesia that is used for procedures of a longer duration. He's in and out and back home within hours. He bounces back quickly. For those procedures, they use an LMA (Laryngeal Mask Airway), so anesthesia doesn't pose the kinds of problems that can occur when the patient has to be intubated.

Andre had an EMG, a procedure where they attached hoops that looked like positive and negative probes on his fingers; one was red and one was black. Then they plugged them into a machine and pulled out another probe that had two points on the end that reminded Jane of some type of zapper. The technician held it in his hand and placed it where the nerves in Andre's hand would be and activated the machine. The machine drew a line that they told him "were like mountains" and they were trying to get "a large mountain." Andre was a real trooper but apparently, as they progressed, the zaps became harder and Jane could see him holding back tears until he finally cried. He was patient, however, and eventually made it through the procedure. The appointment itself probably took the better part of two hours. The test results indicated that he had neuropathies and some weakness more on the left than the right, and that, from all aspects, he had Carpal Tunnel Syndrome. However, as time went by, surgery was not recommended due to fear that anesthesia could compromise his airway.

Andre did have his tonsils and adenoids removed without any complications. Jane remembers telling him he would be able to eat ice cream which he enjoys. He spent the night in ICU, as a precautionary measure, and went home the next day. Whenever Andre has been in the hospital, Jane has been by his side.

When the doctors wanted to perform orthopedic surgery on Andre's legs, they attempted to intubate him about five times. That caused excessive bleeding, so they sent him to ICU for observation, even though they didn't go through with the planned surgery. Once out of ICU, Andre remained in the hospital for a couple of days before returning home.

Jane had a long talk with both the anesthesiologist and the orthopedist after that. They all came to the conclusion that nothing was worth having him trached unless it was a life or death situation. Jane asked the anesthesiologist to write a report for her since they indicated they would not be willing to perform another major surgery requiring intubation without having trach authorization in advance.

Given those circumstances, Jane felt that if they couldn't help his quality of life, there was no need to hinder it. At that stage of his life, even though he had stopped walking independently, it was not worth the risk, since the surgery was not a guarantee. The contractures in his legs and hips are so severe that he really can't straighten up and Jane doesn't want to see him in pain. He really tries sometimes, and it breaks her heart. She knows that he may never be able to walk again and she has accepted that. She would rather he not be able to walk than not be able to talk, and she thinks that a trach would deprive him of that. Even though Andre is soft spoken, Jane feels that there is nothing sweeter than a child who says, "I love you".

Andre had hoped that he would be able to have the surgery so he could walk. Jane had the same hope for him, so it was very disappointing. But she felt that no matter what, she has carried him for all this time and she'll carry him forever.

Joey Nagy

Joey Nagy was only three years old when he had bilateral hip surgery to put his hips in the sockets. A year later, he had a surgery to remove the hardware. They continue to go to a hospital in Chicago where he had surgery on a yearly basis to check his hips; they are still not normal. They learned later that his hips should have been built with shelves to compensate for the lack of bone.

Joey had hand surgery at age six to try and straighten his fingers. He has the clawed hands that are typical with ML. That surgery wasn't as successful as they had hoped, so the second hand was never attempted. When he was ten, and needed surgery for a Carpal Tunnel Release, he had a good attitude. Joey joked about it, saying, "Let's just get it over with."

Joey has a way of laughing off and finding the humor in many hard situations. Linda agreed with him about getting it over with, the sooner the better, so he wouldn't have to spend much time thinking about it.

They did both hands at the same time, with the surgery lasting 1½ hours. The Nagys did have an issue with the hospital when the surgeon, who they were told is one of the best hand surgeons in the country, never came to talk to them either before or after the surgery. When they were able to see Joey after the surgery, they asked for the doctor, only to be told that he had already left. They were shocked when they ended up talking to the intern who Linda had argued with the day before about whether or not Joey had Sanfilippo syndrome. He insisted Joey did because he has type three. Linda informed him that would be correct only if he had MPS. They did learn that the surgery went well. The right hand was worse than the left with severe compression. The ligament that holds all the nerves was extremely thickened. The intern said the doctor never saw anything like it before so he took a piece and sent it to pathology.

While Joey had no complaints of pain in his hands, he did complain about his foot. The nurse had a difficult time finding a vein for the IV that had to put in his foot because of the two hands being operated on. She had to stick him numerous times, and ended up putting it into the area on the inside of the ankle. He was very bruised and swollen there. His recovery from that surgery went so well that they felt it was a "piece of cake" as far as surgeries go.

Joey also had his tonsils and adenoids removed, to treat for sleep apnea, without incidence. When they proceeded to insert tubes in his ears, they found puss in both of them. Linda was very surprised because he had shown no signs he had ear infections. It made her wonder how many times over the years he had infections and they never knew it. Upon release from the hospital, the doctor warned them that the first couple of days were the "honeymoon" period. He was so right! During the night, Joey became congested, which is normal, due to the drainage. Linda went back to work the next day, only to receive a phone call from her mother. Joey was vomiting. He vomited about four times before she finally arrived home, so she called the doctor's office and headed to the emergency room with him. They decided that nausea and vomiting was probably due to the medications he was taking, so they made a change and told him to force fluids. The rest of his recovery went well.

Joey was thirteen when he had bilateral reconstructive hip surgery. Once again, they didn't build them up with shelves. The recovery from that surgery was discussed in chapter three.

Anna James

Anna's first surgery was done when she was only eight-months-old. She needed ear tubes due to the excessive amount of ear infections she had as a baby. She also had to undergo oral surgery at the age of four. Within the period of just a few weeks, Jackie noticed that Anna's back molars had turned black. She had to have several crowns, fillings, and a couple of root canals.

In 2004, Anna's orthopedic surgeon had her tested to confirm that she had Carpal Tunnel Syndrome, after which surgery was scheduled. Jackie said that Anna was nervous about the procedure, but the hospital staff were all wonderful and just fell in love with Anna, who was a little trooper though out the entire ordeal.

The surgeon had put in some pain killing blocks at the nerve, so she had relatively no pain for about twelve hours, but she did experience some discomfort when that wore off. Her parents were very relieved that she had no after effects from the anesthesia and her throat was fine. The procedure was very successful and relatively easy. They were glad that the doctor did both hands at the same time, so she wouldn't have to go through the surgery again a short time later. Anna recovered totally from the carpal tunnel surgery.

In April of 2010, after Anna experienced severe pain in her left knee, she had several visits to the orthopedic surgeon, who wanted to perform surgery as soon as possible. The dates that were open were May 28th (Anna's very last day of grade school) or June 25th (Anna's birthday). When she was asked what she wanted to do, Anna matter-of-factly stated, "Mom, I'm only going to have one more last day at this school and I'm going to have lots more birthdays. You should book me in for my birthday. It's O.K. I'll have lots more birthdays."

Jackie was very touched by Anna's grown up, well thought out decision.

Initially, they expected to do a laparoscopic procedure, but when they went to operate, they discovered a quarter size piece of bone completely broken off of her knee. The lesion in her knee was much more severe than originally thought, so they had to make an incision and drill and pin various parts of the knee. They took that piece of bone out, cleaned, it and then placed it back in by pinning. The surgery went from being a supposed 45 minutes to 2 ½ hours. And of course the recovery time took much longer than they had originally anticipated.

The surgeon later said that the type of fracture she had is usually only caused by high impact running or stress. Normally he would only see that type of thing in an aggressive athlete. And the orthopedic doctor often comments on how Anna's x-rays and MRIs are completely abnormal.

Spencer Gates

In 2005, Spencer had surgery to insert a port (or portacath), which is a small medical appliance that is installed beneath the skin, through which drugs can be injected and blood samples can be drawn many times, usually with less discomfort for the patient than a more typical needle-stick. It was to be used for his Pamidronate infusions. A friend who was a surgical technician at the hospital was able to go into the operating room with Andrea. She held Spencer in her lap until he was out cold, after which she left. The surgery was short and Spencer was in his room in no time. He did well. The anesthesiologist used an LMA (Laryngeal mask airway) that Andrea had learned about through her ISMRD connections. The doctor was not worried about the small airway at all, since the device rests in the back of the mouth and air is pumped down his airway, and there was no tube down his throat.

When Spencer was eleven years old, they met with a surgeon who looked at his X-ray, the MRI of his hips and did a physical exam. He concluded that he wanted to do surgery, specifically Distal and Proximal Femoral Extension Osteotomy. He proposed breaking the bones and putting them back into the correct position so Spencer would be able to stand straight up. Kevin and Andrea decided to think about it while seeking more advice. They scheduled the surgery because there was a May date available, and they knew that they could always cancel. They also met with the anesthesiologist who talked with them and examined Spencer.

A month later they received a call from that hospital stating that they were canceling Spencer's surgery because the anesthesiologist did not feel comfortable with Spencer's airway issues. Andrea was shocked, but made an appointment with a surgeon at a different hospital where Spencer had been seen previously. They felt good about that doctor, who proposed the same type of surgery.

That July Spencer had surgery to straighten his legs by cutting the femur bones, reshaping and putting them back together with plates and pins. They also successfully released the tendons in his groin.

The anesthesiologist took almost two hours to establish an airway using a fiber optic camera, as had been advised by other anesthesiologists who had taken care of ML patients. The surgery itself took about two more hours. That was followed by four hours in Recovery. Andrea and Kevin were anxious when they weren't able to see him for the first two hours after surgery. Further, the doctors kept in the breathing tube as a precaution since it was so difficult place. Unfortunately for Spencer, they kept it in for another day and a half, which frustrated him during the two days in ICU.

When the breathing tube was removed he was very happy to talk, but then he immediately began complaining of pain. He had casts on both legs starting at the extreme upper thigh all the way down to the beginning of his toes. Also there was a wooden pole placed between his legs to keep the tendons stretched out. The casts were to remain for up to six weeks.

After he was moved to a regular room, a few problems surfaced. On Tuesday evening they noticed his hand swelling and the doctors said it was probably from the IV. That seemed strange to them, since the IV in this hand had been removed the previous day. However, the swelling increased, so his right hand was twice the size of his left, and his whole arm swelled up. They were told that it was normal to have swelling after surgery. This made sense, but it had been a week, and it seemed to be getting bigger. When his breathing became strained he was put back on oxygen. They were further concerned that he had been running a fever on and off for a few days.

One week after the surgery Spencer should have been home with his legs propped up watching TV and in little pain. But he was not. On his eighth night in the hospital, he was still complaining of overall discomfort and feeling very discouraged. Four other roommates had stayed for one night after surgery and gone home the next day. The doctors suspected an infection, but his blood and urine samples were negative. The doctors cut holes in his casts at the incision points to see if there were infections and there were not. They started him on heavy antibiotics to see is they could stamp out what they couldn't see. They became concerned with pneumonia and/or a possible blood clot in his lungs, and ordered a CT scan. Spencer continued to go from sleeping comfortably to being extremely agitated with everything and everyone.

Kevin posted a note in the Penguin Café during that time to relate a comical situation that also occurred, "To avoid cursing, Andrea always says, 'God Bless America' instead of 'God xxx it.' Well, Spencer must have been listening, because we must have heard him say, 'God Bless America' at least fifty times a day this past week to us, to the nurses, and the doctors!"

Finally, it was discovered that Spencer had a C-diff (Clostridium Difficile) virus which is highly contagious, and caused by the strong antibiotics, that not only killed the bad bacteria but the good as well, and Spencer's immune system was weakened. He was moved to an isolation room and anyone who entered the room had to wear a gown and gloves. The first medication they put him on seemed to make his swelling worse, so they reasoned he might be allergic to it, and switched him to a different one.

He slept a lot, with labored breathing, which the doctor attributed to his swollen abdomen. He received breathing treatments while he still had a recurrent fever. They saw a positive sign when the swelling in his right hand and arm decreased, and he looked a little better in general appearance

By the end of Spencer's second week post-op, the antibiotics had started to work and he began feeling better. They were able to discontinue the oxygen. The c-diff virus was still causing stomach problems, and he remained in an isolated room. He was able to stay awake for longer periods, and they cleared him to start slowly taking soft foods. He was still receiving IV fluids and he had a GI tube (down his nose directly into the stomach) for other medicines, since this was the easiest way to get oral medicine in him. Then a new problem arose; he developed a pressure ulcer on his tail bone from lying in bed. The wound care specialist said it could go all the way to the bone if not treated immediately. They switched him to a high-tech bed with a mattress filled with sand and moving air so that the surface under his body was always changing. It also heated and cooled as needed. He was monitored closely.

After Spencer spent a total of three weeks in the hospital, they were all relieved to have him home, even though he had a horrible first night there. He simply could not find a comfortable position and was waking every hour. The casts were on both legs and went from the upper thigh down to his toes. He still had the pressure ulcer on his tailbone which made lying on his backside painful, so he had to lie on his side until it was totally healed.

Six weeks after the surgery, Spencer was very happy to have his casts removed. The two pins in each leg that were holding the bones in place were pulled out also, and he commenced to receive Physical Therapy twice per week for the next few months.

Sergio Cardenas

Maria Elena is very thankful that Sergio has needed only a few operations. His quality of life would most likely benefit from some surgical procedures, but it is unadvisable for him to have general anesthesia unless his life is at risk, due to the fact that he has a narrow airway, making it very difficult for him to receive general anesthesia.

He has had surgery five times to insert ear tubes which assist the drainage from his ears, thus helping to avoid ear infections. He has also had his adenoids removed, and a shunt inserted in his head to counteract Hydrocephalus.

Like many other ML children, Sergio has had the Carpal Tunnel Release surgery on both hands. Thankfully, that has been successful.

Sergio has had Laser surgery on his eyes due to Glaucoma. A month after the surgery his eye pressure was lower. Although not completely normal, there was an improvement from what it was before the surgery.

Sergio's doctor said that Sergio's hips are in a very bad shape, but he doesn't want him to have hip replacement surgery unless the boy is confined to bed because of it. Sergio might benefit if he had surgery to repair his faulty heart valves, but the family is trying to avoid that. The situation has presented them with a heart-wrenching dilemma, in that the ML has attacked his body to the point that recovery from such surgery might not even be possible. If a positive recovery could be assured, they would have it done, but the chance of failure is too great.

Zachie Haggett

John and Brenda were not surprised to learn that Zachie needed to have Carpal Tunnel Release surgery on both hands. Five-year-old Zachie did very well with his surgery, although he wasn't happy about the soft casts on his hands when he woke up.

In the summer of 2008, they learned that Zachie had a huge hip fracture. It was very disconcerting, in that he had had many rounds of X-rays before they discovered reason that he had been screaming and crying from pain for at least two months. Brenda cried, right there in the office. Even though they had been back there three times, she felt guilty that they had not pushed for more answers before that. He was scheduled for surgery to put in some hardware to stabilize the hip. For this surgery everyone involved was sure to dot-the-I's and cross-the-T's in preparation. There were many other doctors on standby in case they were needed. The Haggetts felt that the new Anesthesiologist was wonderful. He gave Zachie a sedative to relax him before applying the mask and used the fiberoptic airway. John and Brenda were relieved to receive a call from the OR within twenty minutes to let them know they were done with intubation. Then the surgeon actually clamped the bone back together very tightly. Zachie was out and awake very quickly.

At the first follow-up visit, Zachie had new bone growth around the screw. A month later there was some new bone growth in the area, but the fracture was still visible, so there was still some more growth and healing needed. He did get the thumbs up to start standing on both legs, for as long as he was comfortable.

When Zachie began taking his first independent steps, he used a walker most of the time, but took some steps away from the walker and on his own. He was still very nervous about falling and scared of walking in school. He was excited and happy to be able to walk, even though it was much slower than before. However, his hip did not heal completely. He was unable to stay on his feet for more than thirty minutes a day. In time, it was decided that Zachie would need more surgery on that hip.

As they approached that operation, their usual fear about intubation was put at ease after a long and comforting talk with the two Fiberoptic airway specialists about their plans. They were pleasantly surprised to hear how much prep had gone into the planning for Zach's arrival. Perhaps the three phone calls beforehand to the anesthesiology department paid off. Brenda had made it clear that if she and John were not comfortable with their plan or attitude about Zach and his airway, they would leave before he went into surgery. She was pleased when they walked in the room because of all the anesthesiologists available, she knew they were the two most prepared for her son. They listened intently and discussed all that went wrong previously and put Zachie's parents at as much ease as could be. The intubation took less than twenty minutes. The surgery took a little longer than expected due to the fragility of Zach's bones and the small size of the inside of his hip area. Brenda felt blessed to have a doctor who understands Zachie and his disease equally well, and most of all because he treats him with tremendous love and respect. Before long, the anesthesiologist came out to let them know they had extubated and it went smoothly. But not lost on them was the enormous amount of sweat on his brow. He had clearly been working hard! He admitted that his airway was even worse than he anticipated and that it was the most difficult one he had worked with yet but they went slowly and caringly and got it done. All in all, it was a huge victory in surgeries!

By June 2011, the doctor told Zach yet again that he was still not healed and still could not walk; Zach dropped his head and began to tear up. Brenda found this incredibly painful, knowing how his heart was hurting. He had matured enough to realize what he has lost. He now requires the use of a wheelchair.

CHAPTER EIGHT

EDUCATIONAL CHALLENGES

Parents naturally want their children to have the best life possible. For most parents, that means receiving a quality education. Children with rare diseases usually have unique special needs. Some school systems are well equipped to take the child through the K-12 experience and beyond. Some are not. Most fall somewhere in between. Sometimes parents feel it is necessary to find alternative educational experiences.

Kelley Crompton

Kelley was able to attend public schools throughout all of her growing years, even though she missed many months of actually going to classes. She had frequent colds that caused her to cough very loudly, alarming some of the teachers. They didn't want her there when she sounded like that. And then, she had to miss time for many trips to the clinics where her treatments and studies were being done, as well as when she had surgery. The school provided home tutors during those times, and she was able to graduate from high school by the age of eighteen.

All of Kelley's educational challenges and accomplishments, including the way in which she accomplished earning a college degree, are covered in my previous book, *Kelley's Journey*.

Autumn Tobey

Autumn's school years were often interrupted due to the fact that she had many operations. During those years, she did have one extremely bad teacher who was really mean to her, but Pam didn't know about it at the time because Autumn was too timid and scared of the teacher to even tell her mother. She thought the teacher would be even harder on her if her mother complained. Being a very quiet girl, she managed to plug away and accomplish all she had to do in order to graduate from high school, despite all of the ways in which the disease kept interrupting her education.

She then went on to take college courses. It was only through real grit and determination, which all ML kids appear to have, that she earned her two-year Clerical Certificate in 2001. It took a whopping nine years, with many interruptions for operations, but she refused to give up before she made it. And she made her family very proud of her!

Unfortunately, Autumn never was able to use that achievement in order to find employment, because of all of her health problems. Nevertheless, it gave her a sense of real accomplishment, despite her physical limitations.

Allison Dennis

Alli started pre-school age three, where she tended to parallel play rather than mix with the other children. She attended a mainstream public school (with support), but by the second year, it became very obvious to Trish and Rich that she was not happy. In looking for other options, they found a Catholic special school located about forty minutes away. A huge battle ensued when the educators at the public school, as well as those in the local education department, claimed that that Alli's parents had no right to remove her from mainstream education. They fought the system, asserting that they were more focused on Alli feeling comfortable and fitting in rather than academic outcomes. Trish was under the impression that the teachers in the public sector didn't want to understand why the child had a different learning style. They were not very inventive with learning styles and did not make any changes to suit Alli.

Eventually they won their battle and Alli was happily enrolled in her new school. Yet, the managers of the new school weren't sure Alli could stay, as her IQ and learning levels were higher than their criteria. The school catered to children with mild to moderate intellectual delay. They were not equipped for, and had never had, a student with physical limitations. However, they managed to keep her there on a technicality covered by her reading comprehension. The school administration was very accommodating and even had a ramp and a lift installed for Alli so she could access the upstairs classrooms and library.

Once they moved her to the Catholic school they never had any issues and she blossomed. Alli never looked back and did very well at her new school. Most of her teachers there were wonderful and truly encouraged Alli to be the best she could be.

She was later integrated three days a week into the mainstream Catholic school Mater Dei and continued there right through until the end of her twelfth year. During that year, Alli was chosen as school captain. She also participated in a work experience with their local council and a local newspaper. She had many friends at Mater Dei, but she did not make many friends in the public system.

After graduation, Alli went on to college to study many office skills and take computer courses. The last academic programs she studied was Sign Writing. She is three years along in those studies but has had to stop due to her surgeries. She hopes to be able to go back to college again one day. Alli has never had paid employment due to fatigue, pain and the limitations caused by ML III.

Alli also attended several camps through her school years and later with her respite group. She loved getting away and experiencing new things and they were always very accommodating for her to make sure she had the best experience.

Hayden and Sarah Noble

Hayden attended Kindergarten, but didn't reach his milestones, so at the age of five he went to a Special Needs Kindergarten for six months to get him ready for school. He was five and a half when he went to school, and was mainstreamed with teacher aide help.

Sarah was three years old when she attended a Special Needs Kindergarten, and she was ready for school at age five, when she was mainstreamed with teacher-aid help.

Both children started out in mainstream public schooling. Hayden didn't go to the intermediate school, but stayed in Special Education, and at age thirteen, transferred to a Special Needs Unit attached to a secondary school.

Sarah stayed in mainstream schooling and went to an intermediate school where she had the support of a Special Needs Unit. She transitioned to the same secondary school Special Needs Unit where Hayden was, but for the first year she had some classes in the mainstream and some in the Special Needs Unit. By age fourteen, due to budget cuts in their education system, she stared to fall behind because she had to share a Special Needs teacher. Her parents made the choice to move her completely into the Unit so she could learn at her own pace. However, she did eventually take art classes in the mainstream system and earned a School Certificate in Art and Design and then

6th form Certificate Art and Design, later gaining NECA level 2 in her final year at school.

Both children became what the Ministry of Education calls "Section 99." This means they received a very high level of teaching support. At that time, Hayden was considered a falls-risk due to his paraplegia. He also was able to receive some Physical Therapy at school. They built a special desk for Hayden to use. He is unable to read and has learned very basic sight words, as he is more severely affected than Sarah is. It had been a hard decision for Jenny and Paul to move them into Special Ed, but once they did, they had a safety network and both were able to have work experience, which is something they could never have done in mainstream education.

Their best school years were in the Special Needs Unit at secondary school. They had many friends there, some of whom Hayden still has contact with via cellphone. However, they finished their schooling years when each reached the age of twenty-one.

Sarah was able to participate in a post-secondary education program to complete a certificate in Floristry, but has been unable to secure any jobs due to the nature of her disorder.

Huddy and Sammy Anthony

Liz and Tom Anthony made the decision to home-school Huddy and Sammy for their entire educational experience in order to avoid the many germs and illnesses they would have been exposed to in public school settings. They believed that the boys further benefited from home-schooling in several other ways, in that school subjects were taught to them at their own, slower pace and with total individual attention given to each of them.

They were able to avoid the issue of transportation to school. Nice weather could be enjoyed by doing lessons with chalk on the long, flat asphalt driveway or sitting in the yard. Liz had the help of therapists for determining their individual goals. The boys had plenty of social and community outings and exposure to a large city. They accommodated their particular interests by getting a laptop for Sammy and radios and a DVD player for Huddy.

Liz considers the decision to home-school to be one of the best decisions she ever made. She employed a variety of curriculum, tailoring it to their individual needs and abilities. Every year she kept a journal/assignment book, documenting the activities they engaged in each day, as well as the library books they were reading, and how they were learning. She also recorded who was sick or what doctor they saw and what therapy they did. Every year she listed the books she read to them, numbering over two hundred books of various lengths per year. They didn't learn at a typical grade level, but at their own pace. Hudson could not understand math; Sammy could do sums and subtractions. Both boys learned to read at second-grade level, wrote sentences and wrote their names. They learned to tell analog time, answer cognitive questions about stories, memorize many Bible verses and sing very well. Liz believes that the key to avoiding homeschool burn-out is to adopt a program that *mom* enjoys! They did whatever she thought sounded like fun. Some of her wonderful memories include making peanut butter from scratch, doing a variety of simple art projects that were occupational therapy in disguise, learning to read using homemade reading 'games' besides phonics programs, visiting a friend's 360-chicken farm, a therapy summer camp, boating with the therapists, visiting the fire department with the therapists, reading entire book series out loud (including the Boxcar Children, Hardy Boys, the Narnia series, Pilgrim's Progress for Children, Squanto, and many more) and Physical Education that fit their abilities.

Sammy became very good at using a laptop computer and in 2010 had his own Facebook account. He could use both the mouse and the touch-pad. He was an avid sports fan, following his favorite Wisconsin teams online and on cable. And Sammy had a great sense of humor, which he displayed frequently.

Callie Nagle

Callie had an IEP (Individual Educational Plan) from the time she was in the first grade, due to significant learning difficulties. The Nagles found that hiring an advocate helped them to decipher the IEP. Callie's school work was modified, as well as her homework. Her tests were also modified. Although they were exactly the same as everyone else had, the amount of questions were decreased. In spite of Attention Deficit Disorder plus an oppositional and defiant personality, she has a wonderful sense of humor and is quite insightful.

Throughout her years in school, Callie struggled with self-esteem, mostly because of the way in which ML III affected her body. She tired easily because of her physical restrictions as well as the heart medication she took. Nevertheless, she was able to develop some friendships.

Her parents took her out of Gym class because it was really too difficult for her. She was distressed when she couldn't do what the rest of the kids were doing and felt like she stuck out like a sore thumb. She was able to use that time to get extra help in the school's Learning Center, where there were aides who offered to help the students with their homework.

Debbie learned to look at the big picture of Callie's life when she became tempted to get all worked up over her daughter's school work. She was pleased when Callie wrote a paper in 2005 for one of her classes about a significant event that changed her life. She wrote about having ML III, which was something she had never done. She talked about going to Michigan to meet other kids with ML III and not feeling 'like a fish out of water.' She also said that after going on the trip with Children's Miracle Network, she saw how many people are worse off than she is. Debbie felt that was what sometimes helped her get through the day. It can always be worse!

By the time Callie was ready for high school, her parents found one offering small classes combining "shop" and academics, and had an IEP for every student. Callie was very happy in her new school. Their population included 55% Special Education students. Finally, Callie was in the majority! Even though she had to take time out for operations, she plugged away and completed her courses to graduate from high school. And then she went on to take college courses, doing as much as she has been able to do, even though she has never been a fan of school due to her struggles with learning disabilities. Callie has grown into a wonderful young woman who has made her parents feel very proud of her.

Jennifer Klein

Jennifer has always liked school and has been a good student. By the time she was eight years old and in the third grade, she was unable physically to keep up with other students, but she had no writing or cognitive issues. During the times she was out of school for surgery, the school provided tutoring at her home, and she was always able to stay current with her classwork.

She danced for seven years, but had to give that up when it became too difficult. However, she has played the clarinet since she was eleven years old, with a sound that her mother says is as smooth as butter and quite beautiful. The doctors have been very supportive of this endeavor, as they see it as being a type of therapy for her lungs.

By the time she was in high school, Jenny played the clarinet as she marched in the High School Band for four years. Due to her physical limitations, there were times when the actual marching was very difficult for her, bringing her to her knees and crying from the pain. However, she persisted, because participation was very important to her.

Throughout her life, Jenny has had a friendly, helpful and outgoing personality that has gained her many friends. She has a lot of acceptance of others and is not given to criticism.

After completing high school, Jenny moved on to college, where she is living and working toward a degree, while playing her clarinet in the band. Terri is very relieved that Jenny lives only a short distance away from her, so she can help her out whenever her health presents problems.

Andre Andrews

Jane believes that school issues could probably be discussed indefinitely. When Andre was young he was held back one year to complete a program in a private school, but after that he was able to remain on target. In response to a teacher who wanted to push him ahead, Jane told her she simply needed to take each day as it came, and if he was going to be a superstar, it should be done on his terms and not hers.

When others tried to regulate what he was eating at school, as though they were his primary caretaker, she had to step in and intervene, believing that if her child wanted to eat something in particular or not eat something that he does not care for, the issue should be addressed with her and no one else.

Jane wanted Andre's IEP (Individual Education Plan) to work for her child so that he would receive every resource he needed. Further, while she realizes Special Ed. Teachers have a thankless job of having to cater to individual students, she reasons that's what they get paid to do and that's the job/role they chose to accept.

When Andre was eleven years old, Jane started spending more and more time at the school. She found it to be really draining and difficult to keep Andre from knowing how she really felt about his teacher, although she believed that he was in agreement with her. She didn't feel as bad after speaking to another parent who was having the same problems. She learned that there were advocates that charge a lot of money to help plead a family's case, but all she wanted was for Andre to have a fighting chance at the opportunities that could be made available to him. It was difficult for her to work every day and then help with two hours of homework, plus doing the cooking, housework, and such. But his teacher indicated that Andre's vocabulary wasn't up to where it should be, and he needed to read more. He was required to read thirty books that year (chapter books with six or more chapters and over eighty pages long). However, his teacher had not assisted him with reading any in class and expected that he would read them all at home. Andre couldn't manipulate the pages of a book well, so Jane literally had to hold it while he read. She didn't think that gave the child a fair advantage, and that there should be mechanisms in place for children like him that allow them to be productive and do the same things utilizing different avenues to achieve the same goal.

Andre loves math, but that teacher wouldn't help him to move forward in that area because of a deficiency in reading, which Jane found personally annoying. She wished these types of obstacles and barriers could be moved out of the children's way so they can be the best they can be, but the system is lacking and it can be tough being a parent trying to fix all these things. Jane thought that children ought to have the right to be educated at their own level. She did so much advocating, taking time off constantly for meetings that didn't seem to be accomplishing much, that she often felt drained.

When Andre was in the fifth grade, he was smaller than everyone else in his class. He was still wearing a size two toddler pants, the same size he had been wearing since he was two years old. Fortunately, he had been at the same elementary school, where everyone knew and loved him for five years. But, when he told people outside school that he was twelve, they often stared. Jane felt pretty sensitive about that, while acknowledging that at one point in time we all probably stared at something that looked different to us but tried our best to understand and not stare. She felt that could be attributed to upbringing/home training and sometimes just good old common sense, and saw her position as educating others as best she could, with the hope that people would understand the differences without being critical. Since this was all new to her, she tried to give others the benefit of the doubt. Andre talked about being small and asked if he would get bigger and she tried to reassure him that size isn't everything.

At the age of sixteen, when Andre was in the tenth grade at Phelps Architecture, Construction and Engineering Career High School, he had aspirations of being an architect so he could build homes for people with disabilities. He has always been a math scholar. As part of a Leap Award that year he received a new laptop computer with Skype, and he spends a great deal of time using it. At school, Andre maintained a 3.5 GPA and was a consistent honor roll student, while also participating with the school robotics team.

At the age of eighteen, Andre successfully completed all of his courses and graduated from High School in 2013.

Joey Nagy

Joey Nagy was started in an early intervention program when he was three years old. It was obvious that he was behind the other children. As he became older, the gap became more apparent. At first they said he was developmentally delayed, but as he approached first grade the label became 'learning disabled'. They tried to mainstream him into a regular classroom in first grade, taking him out for special assistance. After seeing some of the work he was bringing home, Linda called for an IEP (Individual Education Plan) meeting. She realized that teachers who have degrees in special education are trained in methods of teaching that works; and for each child, it could be different. She thought that the best placement for Joey would be a Special Ed classroom. There he would be with children with a wide range of abilities and would not feel like a failure. She felt he would not really be getting much out of the other regular classroom, except, of course, socialization. While some parents of special needs children really believe in what they call 'mainstreaming', she feels that there are pros and cons. She wanted Joey to get the best he possibly could from his education. For Joey that meant more individualized attention and repetitive learning, especially with reading and math, which are his hardest subjects.

Linda never regretted pulling him out of the regular classroom. She found it very painful to see her child struggle so much, in every way. She had to work with Joey daily, knowing in her heart that he was trying his very best and that it's just another way this terrible disease has affected him. Joey made wonderful progress in school, making his parents proud of him, as he sometimes he has come up with words of wisdom that totally astound them.

Joey tends to be shy and not one to just mingle in to a new group. Their school program was excellent but they couldn't house the children in the same school for their entire education, so the children move around every couple of years. While it was difficult, Joey tried to take it in stride. However, there was a time when he said he didn't want to grow up. Linda knew it was because he didn't want to have to change schools again. When the time came for him to see his new school, and meet his new teacher, the first thing that came out of his mouth was, "Boy, there are a lot of stairs!"

The regular Physical Education program proved to be too much for Joey when the gym teacher wanted to see how high the students could jump. He experienced a pain in his lower left back right above his hip. School was almost over for that year, so they pulled him out of gym for the rest of the year. His regular doctor had filled out the paperwork advising that he didn't see any restrictions except to let him try everything and do things at his own pace. Linda thought that position needed to be reevaluated.

Once they met with the experts on ML, she learned that ML children cannot be expected to do the same physical things as other children do. They often don't know when to say no, because they want to be able to keep up with other kids. Many teachers, and even doctors, don't realize the limitations caused by this condition. Joey's genetic doctor wrote a beautiful note not only excusing him from gym class, but also explaining the basics of why and what to avoid.

Linda felt much better when they had Joey permanently removed from gym class. The gym teacher at that time decided to make Joey his official helper/drill sergeant. He related a story to Linda about the day that Joey had the two teachers in stitches. The children were running from point A to point B and it was Joey's job to blow the whistle to make them run extra laps if they weren't fast enough. He was laughing very hard and getting a kick out of working the kids while happily blowing the whistle. The teacher said he made their day.

At his last high school before they moved, they had a lot of extracurricular activities for the special needs kids. He had Art Club after school once a week. He was in the Pep Squad which he really enjoyed. They performed next to the cheerleaders. He also was in a national program called Best Buddies. In this program, he was paired with a regular ed. student who took him out once a month to hang out. They also had some group events. Joey enjoyed being part of a peer group of boys where they would meet once a week during school hours to discuss problems. Unfortunately, they moved to a district that doesn't have the funds for such programs. Instead Joey joined Special Olympics, which he had participated in when he was younger.

By May of 2013, Joey had completed all of his courses, and was very proud to have earned his High School Diploma.

Anna James

Anna was able to attend grade school through eighth grade, and for the most part, she was able to keep up with her peers, although there were many activities she couldn't participate in due to her physical limitations. By the time Anna graduated from the eighth grade, Jackie and Bret realized that they were unable to find a good public school with resources to help her in their city, and that a private school providing the extra help Anna needed would be too costly. So they decided they needed to start home-schooling her.

They joined a wonderful home-school group in St. Louis called "Homelink". Anna is able to attend such classes as Literature, Home Economics, Spanish, Art, and Science, as well as classes for special needs students that offer Drama, Money Basics, and Physical Education. They feel this plan has been very successful. She is a couple of years behind the level she would normally be at for most of her classes, so Jackie works with her on History and Science. And, since Anna has been tutored one-on-one in math, her grades have improved considerably.

Jackie owns and operates the London Tea Room in St. Louis, so working at the Tea Room while home-schooling can prove to be a little difficult at times. She does have an office where Anna is able to do her work without supervision all of the time, which has worked out well. Jackie does keep an eye on what she is doing though, as it doesn't take much for her to be distracted!

Spencer Gates

Andrea and Kevin chose a small private school covering Kindergarten through eighth grade, with only one classroom per grade, for Spencer. The principal knows all the families, and not too much slips by her. At Spencer's request, Andrea agreed to speak to his class to let them know about his disease. She felt a little nervous, so she kept it very simple for the kids, telling them that Spencer's body was not the same as others. She started off by talking about how everyone is different; hair, eyes, big, small, different skin tones, etc. She asked them if they knew what a disease was. Most thought it was like a cold or the flu. She explained that it was different than a cold, in that a cold goes away, but Spencer's disease isn't going to go away. He was in the room and very happy to be the center of attention when she spoke to his class. Spencer felt so much better that people knew why he couldn't do some of the things they could do and that his legs hurt him and that he could only raise his arms so high. Just the smile of his face when she completed her presentation made it all worth it.

Spencer really wanted to go to school for the big soccer game between the first and second grades. The whole school watches and cheers them on. He managed to last until the very end of the game, when he just put his head down started crying and could no longer move. Andrea didn't know what to do, finding it sad to watch, since it happened in the middle of the field. It took a lot for her to stop from running out onto the field and rescuing him, but she didn't want to make the situation any worse. A lot of people at the school knew about his condition, so there were teachers and students right there to help him. Andrea thought it was great to see the older students take care of him. She quickly walked to the edge of the field to swoop him up but Kevin beat her to it. Spencer must have convinced himself that he just needed to make it until the end of the game. He is a real trooper, with a tolerance for pain and a stubbornness that she finds amazing.

Even the older kids know who Spencer is. One day when Andrea picked him up from school she could tell he was blue. When she asked him what happened he said that at recess an older boy who he did not know had commented that Spencer was so small. Well, there was a fifth grader close by who told the boy to, "be quiet" and that he was Spencer. Spencer said he was sad and happy at the same time.

Sergio Cardenas

The Cardenas family moved to the United States when Sergio was four years old. In Venezuela children with special needs can't go to public school. There are special schools only for children who are blind, deaf, dumb, or have cerebral palsy. Children with other problems can go private schools, where they can at least try to put the child into a regular class. However, if the child doesn't succeed in a regular class, there are no resources there, so parents need to pay a private teacher at their home or teach the child themselves.

It meant a lot to Sergio's parents that, when they went to register their unaffected children in Texas, they were told that at Sergio's age, he needed to start school also. Maria Elena told them that Sergio had special needs. To her pleasant surprise, they told her that was no excuse - he needed to go to school anyway. Sergio was just learning to talk at that time. He understood Spanish, but he didn't talk in phrases - only simple words. When he began having therapy; physical, speech and occupational, the communication was in English. At school he went directly into an 'English only' classroom, although that school was bilingual. Maria Elena thought that was good, since she didn't want Sergio to be confused with two languages. It was hard enough for him to learn, but at first his mother still spoke to him only in Spanish, as she didn't want to teach him the incorrect pronunciation of English words. After a while, she started to talk to him in both languages, and he helped her to pronounce English words correctly.

Sergio had an excellent Pre-Kindergarten classroom and an excellent Kindergarten/Life Skills Program. At the Elementary school, he started with half of his classes being in Life Skills, with inclusion into the regular classroom. That did not go too well. He started falling behind, and his Special

Ed teacher wasn't very helpful. Unfortunately, he had the same Special Ed teacher for three years and his education went from bad to worse through those years.

He went to a new school for the fourth and fifth grades, and had an excellent teacher for the first semester of the fourth grade. He started to learn again. But then, he forgot everything he learned when he had a new teacher for the remainder of those two grades. Maria Elena tried to get him some help by complaining. She wrote letters and made visits for meetings with the teachers and principals. But all that was accomplished was that the teacher became annoyed, and she took it out on Sergio. She was not kind or patient with him, telling him he had bad manners because he was tired or sleepy all the time at school, which made her think he was bored in her class. However, the bus picked him up over an hour early to take him to a school that was only three miles away. Then teacher complained that Sergio was sleepy every morning, telling him to tell his parents to send him to bed earlier. She seemed to be oblivious to his condition when she sent him alone, without any help to the cafeteria to "learn how to do it by himself." It frightened Maria Elena when he told her that he almost had an accident thanks to the great idea of his teacher.

When she also decided to send him alone to the normal restrooms, he knew that he wouldn't be able to reach the door lock there, so he went instead into the Pre-Kindergarten restrooms where he could more easily reach the toilets and sinks. Another teacher complained to the office about a fifth grade student using the Pre-Kindergarten restroom, indicating that she thought that maybe he wanted to hurt some little students. It was devastating for Sergio to go to the office and be accused of trying to hurt the little kids. He didn't say a word at school, but told his mother the story when he arrived home.

Maria Elena, being very upset by this, wrote a letter to the school lamenting that they always seemed to forget that he had special needs and he wasn't able to do what a normal fifth grade child could do. She pointed out that they knew he was in a Life Skills class, which meant that he needed help 100% of the time. But communication was never satisfactory with that teacher. Maria Elena was left feeling frustrated. She hated to be the mother who sent letters all the time asking for information as to what Sergio was doing during the day, but the teacher never told her anything. She only told Sergio what to bring to school the next day like she would do with a normal child.

Sergio had learned to write his name, and was learning to read and draw when he was in Kindergarten. During the rest of his elementary education he didn't learn very much, only things that could be memorized.

They never taught him to read or write. In the fourth grade he started to read words, and then phrases with eight words, and started writing a few letters and numbers. But, when he got the new teacher, all of that was lost. His parents and siblings have tried to help him, but Maria Elena doesn't want to confuse him, since English is not her primary language. She tried to teach him how to read in Spanish, and he mixed both pronunciations, so she decided to postpone that. She has many conversations with Sergio, covering a wide variety of themes, so he can learn a little of everything and not feel so behind boys his age.

Sergio is able to sing some songs in French and some in Japanese. He learned by watching his favorite Japanese cartoons in Japanese on the internet. He knows some Italian words, too. He loves to sing and dance, and watch videos of rock or pop bands in different languages. Maria Elena loves Italian and Spanish songs, and he loves to sing them with her. That's why she regrets the fact that he can't read. She knows that he is smart enough.

Zachie Haggett

Zachie had a wonderful and supportive extended family at Nursery School for two years. The therapists that worked with him were extremely patient, and his teachers helped him be part of the class and grow. Starting Kindergarten was such an exciting event for him that he behaved like an angel all day. But the day did arrive when Brenda had to leave him at the school doors kicking and screaming after a knock-down, drag-out morning. He cooled off in time, and Brenda assumed that he was testing the water. He was put in time-out for a few minutes one day by one of his teachers and on the way home he told Brenda how much he loved her! She decided that tough love works!

In September of 2006, Zachie entered the first grade. His teachers did try to get him to associate with his peers, but Brenda didn't think his peers noticed him very much. It was a reminder to her that he was different, and it seemed the kids already noticed.

The following year Zachie did pretty well, enjoying his new friends in his class and being at ease with his room where he had the same teacher. He loved school, even if he had a few 'off' days. By the time he was eight years old, he started to show an interest in playing with his peers. His parents saw that as a milestone.

Yet, he had a bit of a rough road when he started back to school the following September, with many new changes happening. He handled some better than others. His parents weren't proud that he had been to see the principal several times in just over two weeks. By the time June rolled around, Brenda was happy to see the school year come to an end. She felt that her little boy had had too many challenges in his very little life in his very little body.

However, by June 2010, Brenda was feeling almost euphoric, looking around and seeing only the good while taking stock in every single blessing that surrounded her. It was Zach's last day of the second grade with the teacher who was lovingly referred to as "Woman" by Zachie. She was one of his first aides from pre-first, and together they became known as the rebels who ruled the school. He lived those entire ten months as happy as Brenda had ever seen him due to the extra support, trust and love of the many people who saw Zachie as a happy, little spit-fire who captivated all who knew him. Brenda and John felt that words could not express their gratitude for the happiness "Woman" brought to Zachie every time she laughed at his silly sense of humor! She also had the ability to know Zachie well enough to notice when something was not right. When he would have an episode with his heart and he wasn't quite verbalizing it yet, she knew something was going on. She knew when he wasn't feeling well. Without her support, Brenda could not have let him go to school with ease. He adored his classmates, his teacher and all of his friends throughout the building, where everybody knew Zachie! His teacher dubbed him "The Man!" He was the most requested boy in class to sit with, to walk with, and to help do projects with. He was accepted, respected, and appreciated by his peers. He seemed to have a connection with every child in his class, small or large. He had something, some common ground with each one that just made him look forward to every day of school. As a parent, Brenda felt that she couldn't ask for more than to have her child be excited to get to school, and she believes that it is rare to find that kind of support.

Zachie's patient and understanding teacher made it her mission to see that Zachie was included in everything his class did, big or small. She never forgot him or overlooked him. A teacher who goes above and beyond to advocate for the children who so often get left behind is a gift from God. Brenda found her to be nothing short of amazing, and feels very blessed to have had her in his life.

CHAPTER NINE

FAMILY LIFE - WHAT IS NORMAL?

People who haven't lived it can't really comprehend the way in which the entire family is impacted when one member of the family has a rare debilitating disease. We can become frustrated with those who judge us and our decisions, as we are try to keep many balls in the air. Some of those balls include the needs of other family members, be they children, siblings or parents. And we have our own share of physical problems as well.

What a balancing act! As though raising children isn't hard enough! We all have our own unique personalities that come into play as we try to live normal daily lives. Yet, our lives are not normal. Stress becomes a constant companion as we try to deal with all the ramifications that come along with raising children who have many medical needs.

Kids can be clever about taking advantage of situations. Affected children might attempt to take advantage of their special needs, and unaffected children can attempt to take advantage of a parent's guilt about not devoting enough time to them. They've got you coming and going!

Once we parents connected through the MPS Society and ISMRD, we found it helpful to share all of our issues on the internet forums. Whenever anyone sends out a request for help, requesting prayers or asking for ideas as to how to deal with a particular situation, the responses are supportive and encouraging. Many of us have tried to find the 'silver linings' on the clouds that appear, and we remind each other to look for those silver linings, too.

In addition to the physical and emotional stress brought about by a rare disease, the financial aspect is usually enormous. There is no insurance or government program that will cover all of the expenses incurred by these families. Very often parents spend countless hours either fighting or pleading with various insurance companies or government agencies in order to obtain help for patients who 'fall through the cracks' because of the nature of the rare disease. Doctors also feel frustrated when they can't find a way to provide a service they believe would benefit the patient, and many of them will do battle along with the parents. These issues are alluded to here and there throughout this book, but we present only a very small part of a very large picture. I hope the reader will bear in mind that every time you get a glimpse of that aspect of dealing with a rare disease, multiply it many times in order to start to understand the stress that is added to a family when trying to deal with the financial frustrations brought on by a rare disease.

Kelley Crompton

Bob and I found that there was a fine line between pampering our child, and giving her he help she truly needed. More than once we explained to Kelley that we didn't want her to whine and complain, yet we didn't want to see her in pain if there was something we could do to help. We didn't want her to become any more dependent than absolutely necessary. Since the disease caused long-term problems, we needed to address these issues time and again.

Kelley wanted to ride a bike just like the other children did, so we let her try. I did have some momentary regrets the day she came home covered with blood, after taking a nasty spill. I put her in the bathtub, clothes and all, to wash off the blood and to assess the situation. Thankfully, her many wounds were superficial. She limped around for a few days and I had to remind myself that many kids take such spills, and if she wanted to ride the bike again, it would have to be all right with me. Her main concern was not the cuts and bruises, but the fact that she smashed her new wristwatch.

When taking jobs outside the home, it was necessary for me to consider the normal needs of my children as well as the many hospitalizations that were demanded by Kelley's disease. We spent countless hours sitting in waiting rooms, emergency rooms and hospital rooms.

Bob and I always had confusing schedules when the children were young and Kelley was in the hospital, trading off our visiting hours with our time at home with the other children. To this day, I am amazed that I never missed a day of the many times she was hospitalized. That was a gift from God!

We managed to think that everything was quite normal when we were between episodes. Perhaps it was a form of denial, but I think it was simply our way of trying to retain sanity. We never expected we would have to be dealing with such out-of-the-ordinary problems when we signed on for the job of parenthood. Basically, we had a typical household for a family of six when the children were growing up and experiencing all the challenges and joys of doing so. We went through the usual chicken pox, poison ivy, sprains, cuts, arguments, talking back, telling lies, blaming each other, and of course, covering up for each other. Typical of that was the time that no one broke the lamp. The baby-sitter did not know how it happened, but *no one* did it!

We realized that a sibling could be giving us problems as they were growing and learning, even if we didn't have a special child. All kids want the attention of their parents, and they will look for negative attention if they aren't getting enough positive attention. We told our children that we would not tolerate any of them being disrespectful to their siblings or parents. We wouldn't let ourselves feel sorry for Kelley or make unnecessary allowances for her. The rest of the world wasn't going to make excuses for her, and the only way she would learn not to make excuses for herself, was for us to be strong and help her to be strong.

When Kelley experienced a severe flare-up in a knee, the orthopedic surgeon who had operated on her hand, took x-rays, and showed them to me, pointing out how the ends of the bones, that were supposed to be rounded, were jagged. He likened them to the Rocky Mountains. He hospitalized her to try a little traction. She was in a room with three other girls, one of whom was a real whiner and complainer. I hoped that Kelley would not adopt this girl's attitude, but she finally did. I walked into her room one afternoon, to be greeted by a daughter with a negative whining attitude. I can be a very empathetic person, but I am of the belief that self-pity hampers recovery. When someone had to deal with as many problems as Kelley did, that kind of thinking needed to be stopped. Just because I believed this, it wasn't easy for me to do what I knew I had to do.

I put my coat back on, and headed for the door saying, "I didn't drive all the way in here today to visit with someone who was going to use up all of her energy on whining, so I'll go home. Maybe you'll be in a better frame of mind when your Dad gets here tonight."

"Wait," she yelled as I reached the door, "I won't do it anymore. Please stay."

I was relieved! We worked on counting blessings.

When she became an adult, Kelley told us that the best thing we did for her was to treat her the same way we treated her siblings. She believed that was what made her strong. She was able to work at various jobs for quite a few years, and to earn a college degree, although the process of obtaining that took many years, because of her many operations. When the day came that she could no longer hold down a job, due to the progression of the disease, she volunteered whenever she was able to. She also spent as much time with her nieces and nephews as she could and loved creating her own versions of personalized stuffed animals to give to them for their birthdays.

During the last five years of her life Kelley required so much care that I was on duty 24/7 to handle all of her medication, transportation, meals, cleaning, washing and dressing. She was so appreciative of all that we did for her, that helping her was really a privilege. I thank God that Bob and I were able to care for her at home until we had to say goodbye.

Autumn Tobey

Autumn's mother, Pam, usually worked twelve to fourteen hours per day for four days a week running her own beauty salon for many years as she has cared for her family. When Autumn started having her physical problems, they had to travel to the hospital in Little Rock, a six hour round-trip, since there were no doctors in their immediate area who knew how to treat Autumn and her disease. Pam felt blessed during the early years that she had wonderful parents who took every trip with them, even providing the car, gas, and food. They also were close to a wonderful family, members of their church, with a daughter the age of Autumn's sister, Michelle. They were more than willing to let Michelle stay with them before school, after school and even overnight, whenever Autumn needed surgery. Although Pam and David didn't have the best insurance, they thanked the Lord that they did have jobs and insurance.

Over the years Pam has had to make many changes in her work schedule. She feels it a blessing that she is able to work for herself and hand-pick her clientele, who have been very willing to work with her, after meeting and getting to know Autumn. They understand that Autumn is her first priority. Pam believes that anyone who can't understand that doesn't really need her as their hairdresser!

By the time Autumn was old enough to work, she was unable to sit for a long time, or stand for a long time. She even she finds it very difficult to get up and down easily. Therefore, she was only able to work at a paying job for a few weeks before her health deteriorated rapidly. Autumn now volunteers at the local elementary school three days a week when she feels up to it.

When Michelle was living in her own apartment, she adopted a precious little Shih Tzu named Max. While Michelle was at church one New Year's Eve, the man that lived in the apartment beneath her fell asleep while smoking and started a fire. Max was badly burned, but the little fella fought so hard to live that Michelle and the vet would not give up on him. Michelle had to move back home temporarily after the fire, and Max needed extensive care when he was finally able to return home from the animal hospital. Autumn learned how to do everything necessary for him to nurse him back to health.

He was terrified when they would leave him all alone, so Michelle knew that she would not be able to move out with Max and leave him alone when she went work. Thus, Autumn offered to move in with her.

As time went by, they became concerned about what would happen to Max when Autumn had to leave the house, too. Max's groomer found Sam (a little Shih Tzu that had been rescued) for Autumn. When Michelle, Autumn and Max went to meet Sam it was love at first sight for all of them. Sam kissed Max repeatedly! And, not only did Autumn and Max have disabilities, but Sam did too - he has only one eye!

Autumn was able to obtain some independence when she lived with Michelle for about three years. Then she required more shoulder surgeries around the time that Michelle got married and moved away, so Autumn had no choice but to move back home with her parents. The dogs moved with her, and the family finds them to be wonderful company for Autumn while Pam and David are at work. The dogs are getting old now, and the family doesn't know how much longer they will be with them, but Pam believes that they were sent by God to help Autumn.

Like most adults, it is hard on Autumn to live at home. She attended her fifteen-year class reunion in 2010, and said that she enjoyed it until everyone told what they were doing with their lives... jobs... marriage... children. Pam finds it heartbreaking to know that is something her daughter will never have.

As the years have passed, their trips for medical treatment have become more frequent. Although it was not easy for Pam when she tried to be of help to her parents as they aged, she is grateful for a wonderful family that stepped up to the plate to help. Her brothers and sisters-in-law helped to take up the slack when she couldn't do as much for her dad and mom as she wanted to when they came to end of their lives.

Pam has found it difficult to keep up the pace she once did, so she has cut down her work to three days a week. She is grateful to God that He has given her the health and ability to work and make enough money to keep a new vehicle and afford the gas to make the trips, sometimes twice a week. She's also thankful that she has the ability to drive in a town like Little Rock, as not everyone can drive in a big city. It is impossible for David to take off work for all these trips, so a sister-in-law tries to go every trip and help with the wheelchair. And, sometimes three sets of ears are better than two when listening to what the doctors have to say.

Allison Dennis

2010 was a very difficult year for the Dennis family. Besides Alli suffering severe spinal problems, her father, Richard, had a stroke. At that time, Trish had to leave full time employment and cut back to working only eight hours a week, since she was needed at home. Then, early in 2013, Rich was rushed to the hospital after he suffered a heart attack. That was followed by major surgery and a long recovery period. Trish continues to take care of her daughter and husband, with great love and without complaint.

Trish describes her daughter as a delightful young woman who loves life! Alli used to read a lot, but has found it more difficult as time has progressed, so she spends a lot of time on the computer and surfing the internet. She really enjoys keeping in touch with others through Facebook. She keeps everyone up to date as to what is going on with her disease, as well as the latest movies she has seen, her latest shopping trips and all of her ups and downs.

When she is feeling well, she is out and about, catching up with friends, listening to music, shopping and going to shows. Alli loves to socialize and has some very good friends who have stuck by her throughout the years. She doesn't have lot of friends, but several very good ones. This makes a huge difference and increases her quality of life.

An animal lover, Alli has two small rescue dogs. Muffin and Missy were both in a very sorry state when the family adopted them, but they have become spoiled and well loved. Alli has a tough time with cold weather, and gets very stiff with loads of aches and pains. However, Muffin and Missy are both very happy and obliging. They love to snuggle down in front of the TV with Alli, giving her loads of cuddles and hugs. They have proven to be a great distraction and a constant source of pleasure. More than anything else, Alli strives to be strong and to take each day as it comes. She has created her own ways of dealing with pain, by focusing on the positive aspects of her life, such as shopping with friends or family members. Even though she often returns home with severe pain, she watches movies and relaxes with her laptop and iPod, while asking others to pray for her. She frequently talks about her desire to overcome the constant pain, and has adopted the moniker of "Brave Warrior Princess." It fits! Alli is an inspiration to all who know her.

In December of 2012, Allie's brother Nathan received a special award for 10 years of volunteering for the disabled. He has a full time job, but still volunteers almost every weekend and sometimes several times during the week. Nathan has also made his family proud.

Hayden and Sarah Noble

Jenny and Paul Noble were young when they married, and they didn't have any real plans beyond what most people think of when starting to raise a family. But, when two of their children were diagnosed with ML III, any dreams and plans that were in their minds, were to be discarded. Paul had to come to terms of never walking his daughter down the aisle. Jenny had to come to grips with Hayden never being able to play a sport. She had expected him to want to play a sport for New Zealand – he always had a bat or a ball in his hands.

They were shocked to be told that their children were two of maybe only sixty children in the world. Jenny wondered why they had been chosen to care for these children, and what the purpose of all this was. She didn't realize at the time the time of diagnosis that she would eventually be in the role of carrying a flag for her children and become an advocate for research for them as well as others with the same and similar diseases.

In the beginning they had no real idea of what they were dealing with, but when the disease was starting to kick in with Hayden, causing falls that could create bad outcomes; they started to protect him while he was in school situations.

Jenny's family didn't seem to understand. Her mother kept comparing Hayden to the other grandchildren, telling her what the differences were – as if Jenny didn't already know! It created a terrible relationship for Jenny with her parents. Her brother, however, has been wonderful, giving strength and support throughout their journey. Jenny's brother and his wife are guardians for Hayden and Sarah should anything happen to Paul and her. Most of their friends have been wonderful, but there were some friends who walked away.

Once Jenny and Paul came to grips with the disease, they started planning the future as to how they would care for Hayden and Sarah for the rest of their lives, planning their housing and where they would eventually live. When they became adults, they wanted their own space. So the family planned a home that has two units; one for them and the other for Paul and Jenny. They live in Tauranga for the warm climate that helps with their pain and joints.

Now that they are no longer in school, they have become somewhat isolated within the community. Paul and Jenny ensure that they have good support services, so they do get out and about. They both play indoor bowls, and have been selected for the National Indoor Bowls Team for Special

Olympics. They have even been head-hunted to try out for the Paralympics in the future. Special Olympics are advocating for indoor bowls to be accepted in the Paralympics. This is huge for them both. They won several gold medals during 2012, having been unbeaten in all games and tournaments.

This disease has opened many doors – doors Jenny would never have seen herself walking through at the time of diagnosis. It has seen her advocate for her children as well as for others around the world. She has given interviews in New Zealand and in the United States, and has presented the Pamidronate data on the scientific stage, being the only non-professional presenting at a scientific meeting. With Paul's help, she serves tirelessly on boards devoted to the advancement of education and research devoted to Lysosomal diseases, while still caring for her own children.

Huddy and Sammy Anthony

There was a time when Huddy was able to walk, but he was all bent over due to contractures. Eventually his main mobility mode became crawling around on his knuckles, even up and down stairs. Both being cute and sociable, Huddy and Sammy could crawl all around, indoors and outdoors, on grass. Huddy has some aspects of autism, causing him to laugh at odd times. He thrives with sameness and lots of praise, and resists transitions much like some of the other ML children, such as sitting on the potty, going to the doctor, getting his hair or nails cut, and going to bed. Liz and Tom understand that is how he deals with things, and they work with it.

Huddy likes to listen to talk radio all day, preferring talk show hosts who use a lot of inflection in their voices. He flaps when excited, repeats a request, and has sensory issues. Huddy likes expecting the same thing. Huddy has a high pain threshold, so there are few complaints from him.

Liz and Tom have both always been able to lift, transfer and carry both boys or push the stroller to appointments, therapies, relatives' houses, church, the zoo, the museum, etc. They've had ten-plus years of Physical, Occupational, and Speech Therapy. They use a disabled parking hang-card in order to have closer parking if it was available.

The disease the Anthony's deal with is serious; it is terminal. The sense of loss is ongoing. The family is always waiting for a crisis to occur. Liz has never been fooled into thinking that her boys could not be "called home to the Lord" at any time.

Before his death on September 11, 2010, Sammy spent a good deal of time at the computer, corresponding with others through Facebook.

Their son Aaron, who is not directly affected by the disease, attends a small Christian school where he has some good friends. His parents let him have as many opportunities as they can without going crazy driving him places. He's involved in activities like competitive chess, flag football and school plays. He is a very good student who spends plenty of time on homework. In his quiet way, he interacted a lot with Sammy who was sociable and intelligent. He talks to Huddy, on Huddy's level. Huddy is absolutely a great young man, and their home and their lives would be lacking without him.

Tom and Liz have done all they can to make their boys' lives more comfortable, exciting or memorable, and they put all else in the Lord's hands.

Callie Nagle

Callie is an only child, and Debbie thinks that might just make things worse for her and her parents, with them putting too much focus on her. She and Richard are both very involved, and usually in agreement as to how to handle situations. Extended family plays a huge part in Callie's life. Of course, being an only child is difficult in itself. She often asked her parents to adopt another child. Debbie did have many regrets that they didn't have another child, but she also wondered how difficult it would be if Callie had a sibling who was able to do everything she couldn't and the resentment that might be there.

During her early years, Callie's behavior presented some difficult times for her parents. She was actually diagnosed with oppositional/defiant disorder. It is very difficult dealing with those issues and the ML issues as well. It's hard enough having a difficult child, but there's a whole other set of feelings and emotions that come with having a child with a chronic illness.

Callie tended to dig in her heels in on everything, and she had many melt-downs. They used to call it "getting stuck" on something. She couldn't move past "no" in most circumstances. So, they started therapy when Callie was about four years old. They used behavior modification and found the reward system worked well, with charts on the fridge to reward good behavior, while they tried to ignore the negative behavior. The therapy helped her to get angry in a good way. Debbie wonders if Callie would have had these same challenges whether she had ML or not. Both she and Richard could see some of the same traits in Callie that they themselves possessed as children.

Since there were many activities Callie wasn't able to do, her parents tried to come up with creative things that she could do. She took drum lessons as well as horseback lessons. Debbie had a scare when, during one of Callie's horseback lessons, she fell off while cantering. Debbie held her breath, watching the scene unfold before her eyes! After riding for eight years, it was the first time Callie fell off. She was very sore, but okay. However, she got up and got back on the horse!

By the time Callie was thirteen, she didn't like the fact that her friends were growing physically, and she wasn't. Debbie found it painful to watch Callie's friends and classmates grow and move on, but she has leaned on her faith in God to help them deal with it. Yet, she feels blessed that they had become very close now that Callie has matured, after the way in which they had previously battled. She believes that it was important to be consistent through all those early years. Debbie found it very helpful for her to talk to a counselor because she had so much guilt as well. The battles they had before school broke her heart! Callie was also later diagnosed with a Sensory Integration disorder which explained why she had major issues with clothes (socks, tags in shirts etc.).

One thing Debbie did when Callie was young was to have older teens visit to "hang-out" with her when Debbie and Richard went out. She was too old for a babysitter, but they didn't like leaving her alone at night. They found some great girls who were good to her. Debbie invited them over to swim in the summer or just pick days for them to visit even if she had nothing going on.

Also, when she was thirteen, Callie was selected to be the Ambassador for Massachusetts by The Children's Miracle Network. One child from each state and thirteen provinces of Canada went to Washington, D.C. where they met President Bush. Then they were off to Disney World. It was an incredible experience for them to be connecting with other parents who have children with some sort of disability/illness, and affliction. It was another reminder to them that others were also dealing successfully with many challenges.

Jenny Klein

When Terri and Walt Klein married they each brought children into their union. Subsequently, their daughter, Jennifer, was born into the mixed family. At that time Lisa was twenty-four, Daniel was ten and Vincent was seven years old.

All seemed to be pretty normal during Jenny's early years in their busy family. A happy and energetic child, by the time Jenny was in school, she took dancing lessons, went ice skating, rode a bicycle, and played soccer, all with enthusiasm. Then the symptoms of her disease started to appear, thus slowly but eventually robbing her of the ability to engage in all of those activities.

Jenny started going through testing to try to determine what was causing her problems at the same time that Terri's mother was spending the final years of her life in the Klein home so they could care for her in a hospice situation. Because she had Amyotrophic Lateral Sclerosis (Lou Gehrig's disease), they didn't want to upset her by telling her that Jenny was having a problem. Terri and Walt also kept the information from the other children, so they wouldn't accidentally tell Terri's mother anything.

What had been a carefree, light and breezy life previous to that time became a life with no peace of mind while they tried to meet all of the needs of the family and juggle jobs. The logistics of meeting all of the physical demands was always on Terri's mind. She was thankful to have some very good friends with whom she could trade off caring for Jenny. She was plunged into mourning the loss of her mother at the same time she was trying to come to grips with the fact that Jenny was having numerous physical problems and would need surgery. She felt that her peace of mind was gone.

Terri and Walt never found an opportunity to actually sit down with the family once they learned the true diagnosis for Jenny was the progressive disease of ML III. The boys were moving on with their own lives, so they were told about Jenny's diagnosis in piecemeal fashion. Daniel was a senior in high school when his grandmother died, and he was going through the process of college acceptance. It seemed like too much was happening all at once. The boys wanted to know what could be done to fix Jenny's problem. It was hard for them to understand that there was no way to actually fix it, but it was something they would be dealing with for the remainder of Jenny's life.

By the time Jennifer was eleven years old, the older children were no longer living at home, so her parents were able to more easily juggle their schedules, as Jenny was like an only child at that time. What had been a schedule of dancing lessons and soccer became visits to doctors and physical therapy. Terri and Walt both had jobs, so they made it work with Terri arranging her hours as a Real Estate Broker around the times that Walt would be at home. Other than that, Terri was able to swap care of the children with friends, especially when Walt's job took him out of town.

Throughout her growing years, Jennifer lived as normal a life as possible, despite numerous problems with infections and bad reactions to many antibiotics. She had ear drum ruptures, strep infections, pneumonia and spinal surgery. She wasn't able to do many of the physical activities she once did, but she participated in Girl Scouts during all of her school years, earning a Gold Award by the time she graduated from high school. Terri became the Camping Leader, as she felt that it made everyone (including parents) in the troop feel better that she was present when they went on trips. Terri gave Jenny as much independence on the campouts as she could.

Despite the fact that she needs to use a scooter to get around most of the time, Jenny's independence was enhanced when she was able to earn a driver's license, and eventually get her own car. Her mother proudly says that she has become a lovely young lady who is an excellent driver, and she is happy that Jenny is able to live at the college that is not far from her home.

Andre Andrews

Andre's siblings are all grown, so that makes him even more special to his mother, Jane. They try to give him as much time as they can, when they can, and Jane understands that they have their own lives to live. There have been occasions where her sister would take the other nieces and nephews out and wouldn't take Andre, which was upsetting to Jane, but she let it go. She had a friend with a handicapped child long before Andre was born, and she saw the struggles her friend had, so she did what she could to help. They lost touch over time and then met again after Andre was born. Her friend tried to return the favor, but with the struggles of her own child, it was a lot of work between the two of them, and sometimes that's the case with family. They try, but sometimes it's a lot of work.

Andre has a deep and abiding faith in God. He and his mother are active members in a Baptist Church where he attended the Sunday school class, which is geared for children with special needs, and he is a church usher. He also participates in a local Saturday bowling league, and travels throughout Maryland and Virginia to compete in tournament play for scholarships. He has participated in various events for children with special needs, like the Starlight Foundation, Dreams for Kids and Kids Enjoy Exercise Now. In addition, he has attended many Nationals baseball games, shows and other holiday activities. Andre enjoys playing video games, and has a Facebook page. He doesn't complain much and realizes that it is others who need to be able to overcome the types of challenges he has in spite of his disability. He gives unselfishly and has a heart of gold.

Jane gives Andre pretty much what he wants to eat, since he is a picky eater. At one of the MPS Society Conferences she attended, the doctors advised the parents that, as long as the food supplies some nutritional value and the child likes it, let him have it. Sometimes Andre says he's not hungry and he won't eat. Jane tries to make sure he has had at least something or she tries to bribe him at that point because she doesn't want him to go to bed and wake up hungry in the middle of the night.

In 2005 life became hectic when Andre had not been able to use his CPAP much because he had constant intervals of cold symptoms. The winter was rough, with Andre being down off and on for three months. It seemed that as soon as he would start to get better, he would relapse again. It didn't help that his teacher was sick and then his aide was sick and probably kids at school and aftercare were, too. Bearing in mind that she is unable to protect him from everything, Jane tries to be cautious with him without becoming overprotective. He still wanted to get out and go to school, so when he managed to get through the week, she doubled the dose on the weekend so he'd be better. She had to give him constant nebulizer treatments. Andre also uses steroids for congestion and/or wheezing. The doctors also recommend a two day dose for him prior to any surgery.

In relating her amazement at the compassion her son displays, Jane says that if she is just carrying him and misses a step, he shows his concern. And when she had surgery on her foot, every time she accidentally hit it Andre would look at her and say, "You okay mommy?"

Despite the fact that their life is anything but normal, Jane proclaims that, "No matter what anyone says, there is nothing more precious than to hear him tell me that he loves me, and hug me tight. Nothing, but nothing else, matters."

Joey Nagy

After receiving the news of Joey's diagnosis, his mother, Linda started to feel like she just didn't fit in anymore, but rather like she was walking around in a kind of daze. She related to Terri Klein's explanation that other people simply can't understand the loss of peace of mind that comes with the diagnosis of a rare disease. She no longer looked at life in same way, but appreciated small things. Observing "normal" healthy families, she wondered if they valued everything they had, while at the same time, she questioned definition of "normal." She resented what seemed to be the trivial complaining of others, and was shocked to hear someone say that they were jealous that the Nagys were for going to Hawaii with the Make-A-Wish Foundation grant! She wondered if they realized exactly what it was that qualified them for the trip to begin with! Were these people jealous of Joey's disease?

Joey was nine years old when he and Linda were in a group called the Sisters of Elegance. It is like the movie Sister Act, where the women dressed up as nuns to sing and dance. They had fun performing at retirement homes, or anywhere else that would have them. Being the youngest member, Joey dressed up as an altar boy. He really enjoyed the group, and all the women really enjoyed him. Although Joey can be shy, he loves to be part of everything.

His brother, Frankie, is four years older than Joey. At times there have been conflicts between the two, as well as between Frankie and his parents, which Linda felt could only be partially tied to normal teenage matters. She believed the rest of it could be due to jealousy over all the attention his younger brother received. Frankie complained that Linda was spoiling Joey and let him get away with everything. However, she felt that she couldn't help but feel overprotective of Joey. Linda believed that he would encounter enough problems outside of the home; at least he should be loved and accepted by everyone at home. But, Frankie didn't want to be bothered by Joey, who was always in the background of everything watching from afar. Joey became frustrated from his lack of attention from his Frankie. His mastery of manipulation caused plenty of fights between the two. Linda became emotionally drained when they had a conflict, as it broke her heart. It also broke Linda's heart when ten-year-old Joey asked her, "Mommy, why do I have to be this way and go to all these doctors?"

At one time Joey didn't complain or even verbalize about how he felt, so Linda tried to encourage him to be open and descriptive of his feelings, in order for them to document and explain his pain. When he went from one extreme to another, she decided that she had a bit of an over-dramatic child on her hands, causing mixed feelings about the situation. She didn't want to discourage him from expressing his pain levels, but on the other hand, Joey had become a little too expressive. One day she received a phone call from the school to pick him up. He said he couldn't breathe and his lungs hurt. She picked him up from school. One minute he was screaming out in pain; the next minute, he was playing normally. She searched to find a happy medium.

It was exciting news for the family when they learned that some of the poems that Joey wrote at the age of twelve, right after his surgery when he was in Children's Memorial Hospital, earned him the honor of being one of a select group of young people chosen as the featured artist in, "The Day the Art Stood Still", the 2008 Snow City Arts Foundation gallery exhibit. His poem entitled "Joey Bear" had been selected. The whole family was invited to attend the opening of the exhibit at the Alfedena Gallery in Chicago. Linda and Frank were very proud of their young man. And Joey was delighted that someone recognized his work.

Anna James

The James family tries to live their lives as normally as possible under the circumstances. The fact that Anna is not able to walk at all, and is confined to her wheelchair, does make it harder. She really depends on others to help her get around. Yet, Jackie says that their family life is good. She does feel empathy for Anna though, as it gets harder and harder for her to get out and do the things that her friends are doing.

It's good that Jackie is able to have her own business, where Anna can spend her time so she doesn't have to be at home alone during the hours that Jackie is working. They also consider themselves lucky that they don't have to travel long distances for Anna's medical care, since they live within ten minutes of one of America's Top Ten Children's Hospitals, where their orthopedic doctor has seen a couple of MPS patients in the past. They feel this is extremely beneficial for them.

Since they bought their house before Anna ended up in her wheelchair full-time, they have to deal with stairs, unfortunately. They hope that they will be able to get a stair-lift installed so Anna can be more independent.

Anna is an avid reader and loves to be in her room reading with her service dog by her side. And, like many of the other youngsters, she does her fair share of posting on Facebook. Her older brother, Peter, has graduated from high school and gone on to college.

Spencer Gates

Andrea, Spencer's mother, says that she and her husband, Kevin, tend to handle things differently. She is Spencer's main caregiver so Kevin pitches in when Sydney, their daughter, needs attention too. She says that raising the two children can have rewards and challenges itself without throwing in a disease to add to the mix. Andrea describes herself as a mother who will do anything she can to make her kids happy.

Their days and weeks are filled with doctor appointments, errands, therapy and a little fun. They liken their lives to rollercoaster rides, with many ups and downs. While Andrea always tries to remember that God gives us only what we can handle, she sometimes she thinks He's gotten her mixed up with someone else. But then He does give her the strength to do whatever is necessary. She is constantly reminding herself to try to balance her time, since she and Kevin usually become over committed with the activities of their children.

When the children were quite young Andrea wondered if Sydney, who is unaffected, complained of pain at times because she was looking for attention. When Sydney asked why God gave Spencer such a bad disease, Andrea explained that God made everyone different and special.

Spencer's sensitivity to the feelings of others was demonstrated when Andrea took the kids for pony rides. They had a great time, but afterwards Spencer said that a metal chain was hurting him during the whole ride. She asked him why he didn't tell the man, and he said that he thought moving the saddle or chain would hurt the horse, and he didn't want that to happen. Sydney was ready to advocate for her brother when she said that he should have told her, and she would have done something about it. Sydney has become Spencer's strongest advocator and protector as she has matured.

If there is an activity or sport Spencer wants to do, and actually can do, his parents let him try it. Baseball is one of those sports. Kevin became "Coach Gates" when there was no one to coach the team Spencer would be on. After he accepted the position, Andrea looked at him and laughed. He looked back at her and laughed, saying that she was to be the Team Mom. The first practice left Spencer in tears, fretting that he was so slow he couldn't do anything the other kids could do. That broke Andrea's heart, but he said he still wanted to play. And he did try to do everything he could, even though he got out every time he got up to bat. He could barely bend over to pick up the ball. At one practice he had on the catcher equipment and fell over. He couldn't get up and was kicking his legs yelling for help, yet Spencer just laughed it off. There was usually a smile on Spencer's face when he was on the field with his friends.

Andrea thought the basketball coach was kidding when he called and said that Spencer had signed up without his parents' knowledge. Andrea didn't want to let him try and didn't do well at the first practice. She was sure that the kids had to find all of her sounds and gasps very distracting. She told Kevin that there was no way she would let him try again. Kevin convinced her to let him call the coach and tell him of Spencer's limitations. At the next practice, Andrea kept her back to the boys. Every time she turned around to look, she saw Spencer shuffling behind the rest of the boys, giving it one hundred percent. A couple of kids walking by saw Spencer and said, "Hey look there's Spencer. He's on the team. That's great! I hope he doesn't get hurt."

Andrea was thankful that she had her sunglasses on, so they wouldn't see her cry. The older kids were looking out for him.

It's in Andrea's nature to say yes to a volunteer position before thinking about it. She became the co-leader of Sydney's Daisy Girl Scouts troop, and Cub-Scout den mother for Spencer's group. Kevin and Andrea assumed that Spencer would not want to, or be able to, move up to Boy Scouts. But when Kevin asked him if he still wanted to do Scouts after those three years were up, he said with a huge excited smile that he wanted to be an Eagle Scout and then an adult leader. He loves scouting so much that they will be involved as long as he wants to.

Talkative and expressive, Sydney is very understanding and helpful to her brother, while she has no problem letting her parents know if she is starting to feel neglected. Before she was old enough to express herself, it was up to Kevin and Andrea to remember to balance their time. It is sometimes hard to do because they have to spend most of their time caring for Spencer. They try to do things with her to make her feel special also. Andrea lives by the words, "Always do the very best you can for today and every day." She hopes Spencer, as well as others, can too.

Sergio Cardenas

After they received Sergio's diagnosis, Maria Elena felt isolated, just as she knows other parents with an ML child have felt. It seemed that other people had put her aside, because her life was different from theirs. In addition to that, she needed to move to another country, very far away from her family, friends, work, comfort, home and even familiar food. Even though the family has had a very hard nine years, Maria Elena doesn't complain. She is very thankful that they are now in the U.S. and feels blessed that they are able to receive much love from the other families with ML children.

However, she still wishes many times that she could have her family nearby, especially her mother and sisters, who could help her with the three young children, and do things like going to school plays, concerts or other special events. When her husband's work has required him to travel away from their home, she is the one who has had to handle everything. So there are still times when she feels very lonely in a strange country.

In her own country Maria Elena had attended a university, with plans to be professional, so in some ways it has felt like a punishment to her to be a stay-at-home mom. Yet she wouldn't change anything they have done in pursuit of a better quality of life for her son.

It broke her heart when Sergio said, "Mom sometimes I wonder how could be the feeling of being a normal child."

Maria Elena thinks that her normal children really are not normal anymore either, because they have all suffered many losses due to Sergio's disease. She believes they have more adult minds than many teenagers. They spend their time helping others and doing volunteer work. One wants to be a doctor and study pain management (Maria Elena playfully says that any coincidence with their reality cannot possibly be true), and the other wants to be a physical therapist (again, no relationship to their reality). Mothers and siblings of special children are special, too. They are special brothers and sisters... and she sees that as not being fair. Maria Elena feels bad about that, as well as guilty. But they do their best to be happy, and to be as "normal" as they can.

The Cardenas family is very involved in the parish life of their Catholic Church. Maria Elena has a very strong faith in God, believing that her faith is the only thing that allows her to keep going while she takes care of her three children and her home, including cleaning, laundry, ironing, taking care of the dog, driving the children to various places and helping them with their activities. When cooking, she makes everything from scratch, as she did in her country of origin.

Sergio likes to participate in the everyday activities that all children enjoy. He likes to play video games and loves to swim even though he does get tired after five to ten minutes. There are many reasons why he tires easily, and that can vary from day to day. Sometimes he has energy rushes during the day; those are the times that Maria Elena cherishes. And then Sergio will have a week when he has a lot of pain, less mobility and trouble with crazy blood pressure, going high and low. He is very small for his age, and he has lost some use of his fingers in his small hands.

Sergio had suffered with petite seizures. At one time he had more than a hundred per day. After using medication for about three years, the seizures started decreasing. Then they appeared again; then disappeared one day, but returned once, and ceased again. From time to time, he displays some involuntary movements, but they are just that, not real epilepsy. The doctor said that they are not sure now that it actually was epilepsy. But they prefer to call those "involuntary movements" or "clinical movements."

Maria Elena believes that fervent prayer to the Blessed Virgin Mary was the most instrumental factor in eliminating Sergio's seizures. Through a friend, and the advocacy of Mother of Carmelo in Italy, she received a blessed medallion that Sergio has used that every day. The doctor can't say if this has had any effect on Sergio, but he has not said that it hasn't helped. Maria Elena believes that divine intervention has been at work.

Zachie Haggett

No matter where Brenda goes, even on a day where she was supposed to be doing something for herself, like getting a haircut, she always ends up talking about Zachie! She tells people that with Zachie, the rules are completely different than that of a "normal" family. In their lives, "Quality over Quantity" is the motto.

The Haggetts have found it hard to be around "normal" children and families because that is not normal for them. She wishes others could understand how difficult and emotionally trying a life with such a rare disease can be. She wants people to know that something as a simple smile from a stranger can make a world of difference in their day.

Zachie has constant respiratory infections. With his hips and legs causing much pain, he becomes extremely tired. He has Physical Therapy five days a week, often sleeping for hours afterward. Zachie's heart function has been strong, despite the fact that the lining of his heart is thicker than normal. Brenda believes that what Zachie lacks in endurance, he more than makes up for in spirit!

Like a number of other others with disabilities, Zachie has a problem with being obsessive about many issues. It is frustrating to have a lack of control over almost everything in his life in addition to dealing with being uncomfortable most of the time.

Even when things are going well, Brenda feels like they often are "waiting for the other shoe to drop." She believes that God knows what lies ahead for them, but it is tiring to try to push back the fear and loss. The facts are that ML II/III kids do not live long lives. They want to hold onto hope for treatment, but they also have to be realistic and they have to decide what they are willing to put him through while he is here with them, evaluating what may be "wants" for them and absolute necessities for Zach.

It was very disappointing when they realized that Zachie would have to use his wheelchair permanently. Knowing the day would come didn't make it any easier to handle, since every little thing he worked so hard to achieve was being taken away, a piece at a time.

As a mother, Brenda only wants to see the good things happen for her son, but the reality of the way the disease has ravaged Zachie's body is a reality that has been very difficult to deal with. As she was dressing him for school one day he asked, "Mommy, when am I going to be normal again? When am I going to not hurt all the time?"

It became clear to her that as much as his normal has always been this way, with hurting and hospital visits, it is now wearing thin on his immense heart and courage. Hearing those words and seeing that look on his face caused her much pain and sorrow. She had to remind herself that God does not purposely give children pain and suffering but continues to guide us through these times with continued strength and courage. Yet, it is no easy task looking into those big beautiful brown eyes filled with sadness and defeat.

Within the group of parents who shared their thoughts and feelings in the Penguin Café, there has been a discussion of finding the "silver linings" on every cloud.

Brenda writes:

"The Silver lining here is....

Some people miss the days of cuddling with their twelve-year-old son, I enjoy every single moment of cuddling thus far.

Some people miss the days of carrying their child to and from bed but I do it every day still while stealing a kiss all the way.

Some people wish they could go back to the days of endless 'I love you's' but I notice and revel in every single one.

Some miss little tiny feet in the house, I still get to tickle those fat little tootsies every chance I get!

Some long for an occasional compliment to come their way from their twelve-year-old going on too-cool-for-you while I am blessed to be told that I am "the best mom he's ever had" at least once a day.

For every tantrum that is way beyond the terrible twos I am reminded of the very fact that I still have tantrums to deal with!

For the one hundredth trip for chicken nuggets I try to remember when he could barely keep down liquid Boost.

For all of the future goals the world sets for tweens and teens these days, I am grateful for all of the goals we have already achieved to get to today.

For all who can't wait for the tomorrows; I am grateful to have been blessed with Today and I will revel in Today all day long because we still have him today!

Thank You, Lord for giving me this blessing of a small but mighty little being who continues to raise me.

And also thank you to that little birdie who always reminds me to look at the silver lining."

CHAPTER TEN

CLOSE CALLS

In dealing with the many aspects of the problems caused by our rare diseases, some of our families have encountered nerve-racking emergency circumstances, when emergency technicians and critical care units have performed heroically, to our great relief.

Kelley Crompton

I don't know how many times that Kelley visited an emergency department throughout her life, at one hospital or another, for one reason or another. Most of those visits were due to breathing problems that couldn't be handled at home often resulting in her being admitted for treatment, but there were other occasions as well. I always carried copies of 'Kelley's Medical Resume' with me, since I couldn't possibly remember all of her medical history without it. Every procedure and/or hospitalization was listed, as well as her allergies and current medications. The names and phone numbers of the doctors currently treating her were also listed. Everyone I ever handed it to responded with surprise and remarked that they wished everyone would do that. It was also a helpful tool when filling out those forms at the first visit to a new doctor. There was never enough room on those few lines allowed for the patient's history. There was always enough room for "see attached."

In 2003 we looked forward to attending the MPS Conference in St. Louis because it was the first time that we knew that another ML III family would be attending. We planned to meet up with Brent, Jackie, and Anna James at the conference.

When we arrived at the second motel on our way to St. Louis from New Hampshire, we checked in, and Kelley and I went to our room while Bob started to bring in our luggage from the car. Kelley was in the bathroom when I heard a thud followed by whimpering, "Oh, no! Oh, no!"

I ran to the bathroom and opened the door to discover my daughter in a very strange position on the floor. What a sight! Her right artificial hip had dislocated, causing her severe pain. We were quite surprised. She never had any problems with that hip in the fifteen years since it had been replaced. Her leg was at such an unusual angle, that I knew we shouldn't move her. I called the front desk to request an ambulance, quickly explaining what had happened. Then I grabbed a spread and blanket off a bed, placing the blanket under her as best as I could, and the spread over her while we waited for the ambulance. When Bob returned to the room there was a motel attendant there too. I was sure they were worried about their guest getting hurt on the premises. I quickly explained to my shocked husband what had taken place.

How fortunate we were that the hospital was a short distance from our motel; the EMT team arrived quickly. Since the bathroom was small, it was quite difficult for them to extricate her without making the situation any worse.

Bob and I followed the ambulance to the hospital. The entire time, Kelley was asking, "Why now?"

However, she wasn't asking, "Why me?"

Her visions of spending days in a strange hospital in Indiana and missing the conference caused her concern. I was less concerned about missing the conference, but quite concerned that we were so far away from her doctors. Who knew if the doctors at the local hospital knew anything about tracheas like hers, if she would need surgery?

It seemed to take forever to get them to give her some pain medication. At that, it didn't help much. When the orthopedic surgeon showed up (after his office hours were over), I asked if they could do a 'closed reduction' rather than have to take her into surgery. He had looked at the x-rays that had been taken, and assured me they could. That was a major relief!

Then he told the attendants to take her to a room that had a door on it. And after we were settled in there, he told me to leave, saying that he would get me when they were finished. I asked if she was in that room because she was going to holler, and he didn't want me to hear it. He confirmed such, so I left the room saying, "Kelley, he just gave you permission to yell."

Once her hip was in place, she was exhausted. Nevertheless, she quickly asked the doctor if she could still go on to our next destination. He replied that she could, because she would need to use crutches for the next month, no matter where she was. She was instructed to see her orthopedic surgeon after returning home. It was unusual for the hip to displace, but it happened after she had been very sick, and had not done any exercising for some time. Her muscle tone was poor, so all she had to do was to move into the wrong position.

When we returned to the motel, we were asked to complete an incident report. I told them not to worry; it wasn't their fault - we had no plans to sue, and I would sign whatever they needed me to. We just wanted to get some sleep and move on.

It was a few years later when Kelley's orthopedic surgeon disappointed us by telling us that he would no longer be performing surgery. We had complete confidence in him and the anesthesiologist who always took care of Kelley during her orthopedic surgeries. Since she would need a new surgeon, we decided to see if there were doctors who could handle her trachea at the medical center where she was already being seen in the Pain Clinic and receiving Pamidronate infusions. We reasoned that it would be good to have all of her major care under one roof. The thoracic surgeon I contacted seemed to be reluctant to take on someone with her diagnosis and medical history, but I think he liked her when we did meet with him. He told us that they had a very capable anesthesiologist on staff who would be called in when it was time for Kelley's t-tube to be changed again, and also when she would need orthopedic surgery. Thus, we were able to establish a relationship with a few more doctors at that medical center before then next situation arose.

Kelley started to have a problem with some liquid getting into her lungs when she swallowed, so she had extensive testing at the medical center to determine why she was aspirating. After receiving inconclusive results, she agreed to be sustained with a feeding tube, at least for a while. Her trachea and larynx had become very swollen during that time, and she was presented with the possibility of a laryngectomy. She did not complain, but said, "Well, I could live without eating, and I could even live without speaking, but I know I could not live without breathing. So, if I have to live with those other things, at least I will still be able to breathe."

She had the will to go on, and the ability to accept whatever came her way. When the nurses apologized for poking her time after time to set up intravenous lines, she told them that it is not their fault that she was a 'hard stick.' Never one to feel sorry for herself, Kelley showed no anger or sadness, but managed to make the best of any situation, all the while trusting God.

Ten days after the feeding tube was surgically inserted we went back to the medical center, as it was necessary for them to remove some sutures. All seemed to go well, and we returned home and set up her feeding to run through the night, which is what she preferred, so she wouldn't have to be tied down to it during the day.

Kelley's screaming awakened me at two o'clock in the morning. Her abdomen was distended and she was in tremendous pain. I quickly disconnected the link to the infusion and wasted no time in getting her to the ER at our local hospital, where X-rays showed that the feeding tube had slipped out of her intestines, and the fluid was collecting in her abdominal cavity, due not only to the feeding but also to the gastric juices escaping through the hole that had been created. The ER doctor ordered a strong pain medication intravenously, explaining that peritonitis was one of the most acute kinds of pain a person could experience. I had never seen her react to the pain by thrashing the way she did at that time. It took four of us to hold her down so they could get an IV started. And that seemed to take forever due to her resistant veins. The doctor asked who her surgeon was, and I explained that there was no one locally who would take her into surgery because of her airway. I gave him the name of a few of the doctors at the medical center, after which arrangements were made for her to be transported there via ambulance.

Shortly after we met up with Kelley in the emergency department at the medical center, she said to me, "Will you please take this penguin that is in my hand?"

Since this took place shortly after we had made connection with the 'Penguins' of ISMRD, I thought she was hallucinating from the drugs they had given her. However, she opened her hand to show me a tiny stuffed penguin! I could barely believe it. She said that the ambulance EMT gave it to her. I asked her if he knew about our connection to the ISMRD Penguins, and she said not, but she thought it was something that they had from fund-raising they had been doing. I chose to believe that it was God's way of reminding us that we had a lot of good friends out there!

The ER doctors at the medical center asked for consults with numerous doctors. They were rightly concerned about Kelley's airway, but they had to take her into surgery, because her abdomen was becoming more distended by the hour. She was running a fever, her white blood cell count was up, and her pain was excruciating. The doctors acknowledged the severity of her pain, and gave her as much medication as they could. Two of the doctors who had recently treated her were called in. We were glad to see them. Bob and I listened intently to their detailed discussion as to how they would treat her airway while the general surgeon performed an exploratory laparotomy to take care of her peritonitis.

The surgery went well, and was finished by four o'clock in the afternoon. They cleaned out her abdominal cavity, repositioned the feeding tube, and even repaired an umbilical hernia. Normally she would have gone to a regular room after the recovery room, but they felt that the history of her airway demanded caution. She was sent to the critical care department, where she was heavily sedated and placed on a ventilator.

The following day, they weaned her off the ventilator. There were numerous attachments and bags hanging for IVs to pump three different kinds of antibiotics into her. Abdominal surgery required those precautions. They also started feeding her via the tube again, pacing the infusion slowly. She did need to be suctioned fairly often, as she didn't have enough power to cough anything up - not to mention the pain caused by coughing.

She left the hospital with antibiotics to keep infections at bay. The first week home was taxing. As she was healing from the surgery, her trachea was still irritated, making frequent suctioning necessary. We were on a schedule where all day long, every two hours, we had to either give a medication that had been crushed and dissolved (as much as possible) in water so it could go into her tube, or she needed her tube flushed with water. She wanted to have the food running in overnight, so she could have the freedom to move around at will during the day, but the benefits of one of her medications was that it could be decreased if she had it too close to her liquid feeding. Therefore, all of that was on-again, off-again, and some of the food needed to be infused during the day. The last medication of the day was at ten o'clock at night, after which we set up the feeding again, and the feeding was done in time for the first medication at six o'clock in the morning. Our aim was for us to get a good break for sleep, if she didn't wake up requiring suctioning, which she did more frequently than we had hoped.

A month after the hospital discharge, Kelley had a new swallow test followed by a visit with the ENT doctor. The results of the test showed that she was still aspirating clear liquid, but not solids. So the doctor said she could safely start to slowly get back to a normal diet, with hydration being given through her feeding tube. Our impression was that the swallow test results did not show a completely clear picture of the reason for aspiration problems Kelley had been having. We did understand that liquid was being aspirated all of the time, so Kelley decided to keep obtaining hydration through the feeding tube, but to try eating, taking the process slowly while still having supplementary feedings through the tube. However, whenever she did eat, it seemed to trigger something that caused her to cough. That must have been quite abrasive to her trachea, as she made more secretions, most times requiring suctioning. Therefore, I kept the suction equipment handy when she ate. It was very helpful to have her take her medications by mouth, since, when they were crushed, it was sometimes very difficult to get them through the tube. We did have a few trips to the ER to have the feeding tube changed when the medication got stuck.

One of the doctors at the medical center suggested that Kelley might want to see a different doctor in the big city about having stents placed in her trachea. I looked up that doctor's information on the internet and gave Kelley a report. She wasn't at all enthusiastic about exploring the idea of going back to the city. We agreed that her tracheal problems would have to be handled at the medical center. We had become very comfortable there, and believed it to be the right place for Kelley's care.

During our next visit with the ENT doctor in early December, one of the things he mentioned was the possibility of a fistula. We discussed it only briefly. Kelley did not want any more testing that day, saying that her larynx was so sore from the swallow test that she didn't want to be scoped. After our return home, I started doing some research, learning that a tracheoesophageal fistula is a situation in which an opening occurred between the trachea and esophagus. In such a situation, some food/liquid traveling down the esophagus crosses over to the trachea, thus finding its way into the bronchial tubes and lungs. The symptoms Kelley had at that time appeared to fit that picture. I didn't know quite how that diagnosis was determined, especially when one had a tube in their trachea. Since she was planning to have the t-tube evaluated and changed sometime in January, I hoped that she could wait until then to have it evaluated. Yet, I wasn't sure how healthy it was for her to wait that long, even though she really didn't want to go back to the medical center before then. Her coughing, while better than it was post-op, was still problematic. As we look back on that time, we can only recall it as a blur of day turning into night, and back to day, with no end to the problems in sight.

On December 12th, 2006, Kelley had gone downstairs to get ready for breakfast while I was upstairs getting dressed. Suddenly I heard her cry out, "Help!"

I ran down the stairs to find her sitting in a chair, in severe distress, trying to catch her breath. I started to suction her but it didn't seem to help. The color started to drain from her face, and she was becoming panicky. I knew my efforts weren't going to solve the problem, so I asked Bob to call 911. He stayed on the phone with the emergency operator while I kept trying to suction Kelley. I wasn't going to give up, but I didn't seem to be making any progress as her lips started to take on a blue coloring. I kept glancing out the window, realizing that this was first time that I was afraid the 911 responders would not arrive in time. I felt powerless as she gasped for breath and managed to whisper, "I'm going to die."

"Not today!" I demanded as she fell backward in the chair, passing out.

Earlier that week, I was suctioning her in the middle of the night when she said, "I'm sorry I had to get you up Mom, but I'm praying for a miracle." My heart broke as I fought back tears. Certainly, no one would have faulted her if she had given up, being already exhausted from the four months of hospitalizations, surgeries, suctioning, feeding tubes, medications and interrupted sleep. I couldn't imagine that there would be an answer this time.

When the emergency team arrived, after what felt like an eternity, I yelled, "Oxygen."

Once they started the oxygen, administered through the arm of her t-tube, she started to respond. I quickly informed them of all that had been going on for that past four months. They stabilized her and headed for the local ER. We followed.

I explained to the doctors at the hospital that I had found it necessary to switch to a smaller size catheter to suction her, because I had been having trouble getting the larger one down the tube. They called in a pulmonologist and an ENT doctor, both of whom were familiar with Kelley. After doing a scope, they found that significant secretions had adhered to the t-tube, nearly obstructing passage. Those doctors worked feverishly to keep Kelley breathing for many hours. Clearly, the t-tube needed to be changed, but they were not equipped to do so there. They spent many hours trying to reduce the secretions, reduce Kelley's anxiety, and make plans as to what they would do if they lost her airway completely. After many tries, they found the right amount of medication to keep Kelley's high anxiety down without reducing her ability to breathe. They made plans to obtain an airlift. Time after time, the doctors would get everything under control, and go back to their respective duties until they were paged again, and they came running. It was a very tense situation, the likes of which we had never encountered in the emergency department previously. We felt it to be a blessing that Kelley was not too aware of all that was happening. When a call came for me from the admitting office at the medical center, I asked if that meant they had a bed for her, and was told that indeed they did. Then we heard the notice over the PA to prepare for the arrival of the helicopter, and a nurse ran in saying to us, "That's for you."

We felt relieved by that, but our anxiety was still high. During that time a nurse that we had not seen previously that day arrived on the scene and sort of 'hung out' with us. It crossed my mind that the same woman had appeared when we waited for the ambulance to take Kelley to the medical center a few months before that, when she had the peritonitis. I assume that her job was to make sure the family was not ignored. Although I didn't see any need for it at that hospital. They kept us completely involved at all times. Yet, I do realize that some people might need to be comforted at such a time.

The helicopter crew gave us confidence that they could take care of our daughter as she traveled to her next destination. We were impressed by the way they took notes from the doctors. They had large strips of white tape on their pants at front of their legs, where they wrote the notes as the doctors reported. No need to worry about losing any paperwork!

That extra nurse accompanied us out to the lot where we watched them load Kelley into the helicopter and take flight. We knew she would be at the destination long before we would. We could only pray that they could keep her alive during the fifteen minute trip.

As we drove to the medical center, we discussed how well the doctors and nurses at our local hospital handled the situation, and that we knew we had made the right decision in making the change from the big city to the medical center. Upon our arrival there, we discovered that the capable and professional med-flight crew had delivered Kelley alive, and the staff at the medical center had everything under control. The variety of equipment they were using allowed the patient to rest while they made arrangements to take her into surgery in the morning. She was pretty much 'out of it' when we kissed her goodnight and went home for some sleep.

When we arrived at the medical center the next morning, I asked one of the nurses to have the doctor check Kelley's trachea for a fistula when they were changing her tube. I reasoned that she would be under anesthesia already, so it would save another trip to the OR to check that out at another time. She had me write a note to put on the front of Kelley's chart, so the doctor would be sure to know of my request. We made our way to the surgery waiting room to register with the receptionist and start our wait. Shortly after that, a resident doctor came to talk to us. He said that the attending surgeon had received my note, but didn't think he would have the time to do any more than change the t-tube at that time, so she could breathe again on her own. We understood. It was emergency surgery, and they had to fit her into the schedule.

It was a while later when the surgeon appeared in the waiting room and motioned us into the private room, closing the door behind us. He said that she was still alive, even though, "She tried to die on us." He went on to explain that they had removed the old tube and inserted the new one, but something wasn't right. He assured us that the two other surgeons, who were working with him, would do whatever was necessary to keep her breathing. We knew enough about tracheal anatomy to understand what he meant when he said that he inserted the new tube and looked through the scope for the positioning, he couldn't see the carina. That is the name given to the base of the trachea, before it splits to form the two bronchial branches. He said he didn't know what he was looking at, but he wanted to give us a progress report. Then he departed, leaving us wondering just what it all meant.

The next time he appeared, it was to tell us, "Kelley keeps trying to die on us." He went on to explain that they realized they were really looking at the esophagus, because the t-tube had found its way from the trachea into the esophagus. In other words, they had found the fistula I had been wondering about. Apparently a tear in her trachea had created a canal for the tube to pass through. We didn't know when the problem had started, but it was probably after the last time she had the tube changed at the other hospital, a year before that. What it meant was that, when she would drink or eat, some of the substance had been getting into her bronchial tubes and lungs, so she had to cough it up. It appeared the fistula had been growing until it reached a critical point. After many attempts to get the t-tube to stay in her trachea, they decided they would have to try some other approach. Kelley was stabilized and taken to Intensive Care, to be kept heavily sedated on life support until such time as they would be able to take her back to surgery - once they worked out a game plan.

We had no idea if the doctors would be successful in solving the problem. When we visited with her in ICU, she didn't have any knowledge of our presence. There were so many people caring for her, that we weren't concerned about her safety during that time. However, we were very concerned about her next trip to the surgical suite. We had heard so many times about Kelley's delicate trachea that we wondered if they would find a solution.

Kelley's brother, David, and sister, Peggy, waited with us that Friday when they took Kelley back into surgery. While sitting in the surgery waiting room, I thanked God that we had found the doctors at this medical center. I took out the piece of paper I called 'Kelley's Medical Resume' to count the times she had been to the OR. This would be number forty-five. Could she make it again? I thought about the fact that she had trained to be a teacher, and although she never had a classroom of her own, she indeed was a teacher. Many people had told me that they had learned about acceptance and grace from her. Would we all be as accepting of God's will if this was to be the time He chose to take Kelley away from us?

We were anxiously waiting to hear from the surgeon, when he emerged to tell us that she was still alive. He went on to say that that they had decided to try placing a stent in her trachea to keep it open, and to block the fistula, so food wouldn't be able to cross from the esophagus to the trachea. However, tracheal stents had not been used in that facility previously, and they only had one stent the appropriate size in the hospital. It proved to be insufficient.

Thus, it was necessary to send her back to Intensive Care and keep her on life support for a few more days, while they waited for more stents to be delivered. They gave us no assurances that the procedure would work, but there really was no other option. The cardio-thoracic surgeon told us that he had called numerous other surgeons around the country to see if any of them had any better ideas. He also joked with us that he now knew why he was so reluctant to answer my original letter to him. Her previous surgeon from the big city (who had pretty much given up on her that past January) simply wished him luck!

We realized that the doctor who had approached us about taking Kelley back to the big city (to see the doctor who used stents) was the one who came up with the idea to try them in her. It was easy to imagine that these doctors would be disinclined to try something they had not done before, but they were presented a difficult situation, and they determined to do their best to deal with it.

Monday arrived, as we tried to prepare ourselves. Kelley was facing very delicate surgery, and we were thankful that many people were praying for her. The idea of Christmas without her was unthinkable. They took her back to the OR to see if they could finish the job in the way they hoped to, advising us that this kind of stent was pretty new for use in the trachea, but it was worth a try, because there didn't appear to be any other solution. They reminded us again about her very unusual anatomy!

The woman on duty in the surgical waiting room knew about the situation pretty well by that time. She kept tabs for us while we waited. I think she was as happy as we were when the surgeon emerged to tell us that Kelley was breathing on her own. We cried tears of joy! The surgeons approach had worked! Two stents were placed in the lower part of her trachea. Being concerned about the swelling in her larynx, they didn't want to place a third stent at that time, thus blocking access to airway if her larynx closed up. The solution to that was to place a 'button' where her t-tube had previously emerged through her neck. The plan was for her to continue with the tube feedings to give her fistula a chance to heal. We were told not to suction her unless absolutely necessary while the stents were settling in. That seemed like a novel idea!

Since the diameter of the stents was much larger than her t-tube was, they expected her to have easier time breathing. The surgeon said that the stents were like a 6-lane highway compared to the tube she had before. He also said that this type of solution was not available when she first got her tube. It appeared that the surgery was successful. She was breathing on her own, supplemented by some oxygen, when we left her. Her body was no longer making secretions like it had been, so she didn't need to be suctioned all the time.

Within a few days, Kelley was in a regular room, without any supplemental oxygen. The staff at that hospital gave us the best Christmas gift when we brought her home on December 24! With no complaints about the feeding tube, Kelley was happy to be able to breathe easily and sleep through the night. I assumed that God believed that she still had more to teach us about perseverance and hope. Kelley's miracle had been delivered. We called it our Christmas Miracle, and asked God to bless all of the wonderful people who had helped her through that crisis.

Allison Dennis

During 2009 and 2010, as Alli's lower back pain was worsening, Trish took her to see many spinal surgeons, all of whom claimed that her condition was not severe enough for them to operate at that time. As Alli's pain progressed from bad to worse, Trish decided to take her to a major hospital and ask Alli's geneticist (who was knowledgeable of the way in which both MPS/ML could affect a patient) if he would please help them. He looked at Alli's scans and wiped tears from his eyes, as he told them that the problem is that many doctors just don't realize and won't listen (even to him) as to how bad her disease really is. During this time they experienced problems with the Pain Clinic doctor not wanting to understand the source of Alli's pain. After a heated discussion, Trish asked for Alli to be referred to someone else. She was referred to a rheumatologist who did listen and was helpful with prescribing pain medications. On a regular visit to him, Trish mentioned Alli's worsening pain and limitations. After examining her, he had her admitted to the hospital immediately, saying that there were obvious signs of spinal compression. Alli was admitted and went through many tests, examinations and X-rays. During this time her legs and arms were becoming weaker and her bladder stopped functioning. She was in the hospital for six days, when their neurologist told them that there was nothing that they could do. She said that they would send her home with a catheter, and they should make the most of what time she had left. Trish was horrified!

Remembering that Hayden Noble had been through spinal surgeries, Trish contacted Jenny Noble in NZ to see if she could help. Jenny asked her to have the doctor call her children's specialist in NZ. Trish begged the Neurologist to do so, and she agreed. Later that afternoon a Neurosurgeon and his team visited Alli. He had previously looked at her scans and said there wasn't anything he could do. Now he took another look at Alli and told Trish that her daughter had to go into surgery immediately. Within an hour she was on the operating table. Eight very long anxiety producing hours later she emerged from surgery at two o'clock in the morning. Trish was very relieved when the Neurosurgeon told her that Alli was OK, with a seven inch rod in her spine, held in place by screws. He explained that the reason no one realized the extent of the problem was because the spinal cord was twisted and could only be seen in flexion. That was when they learned that all of the scans that had been done weren't showing the full extent of the damage of the spinal compression. Alli was carefully monitored for a few days in Intensive Care after the surgery. Ten days later, she was well enough to return home, with the hope that any further surgery wouldn't have to be done under such extreme circumstances.

Hayden and Sarah Noble

The Nobles have always been told to carefully manage the spines in Hayden and Sarah, but by the time the doctors realized that they had spinal issues going on, both Hayden and Sarah were getting into serious situations. They were experiencing headaches, back pain, pins and needles in their arms and legs and chronic tiredness. These symptoms can often mean other things as well, so when they finally reacted to the symptoms, they were seeing a decline in quality of life in both Hayden and Sarah.

Jenny noticed that the two of them became very good at hiding what was going on. They didn't want to worry their parents, as they saw the struggles they had in attempting to get the right care for them in Adult Services. They found it extremely distressing in trying to get tests done and get results in, as well as trying to find someone who would take what was happening to Hayden and Sarah seriously. It took almost two years to find the right person to take on the very complex necessary operations. They were pushed from one doctor to another. Some said a Pediatric team needed to be involved. Pediatric Teams said it was Adult Services responsibility to care for Hayden and Sarah. One team of Adult Services doctors said the case was too complex and had too high a risk, so they should go back to Pediatrics. This is one of the complex issues they face in New Zealand in obtaining coordinated care for patients with complex conditions. They eventually found a private doctor who agreed that Hayden had a very serious case of instability and cord compression. Yet, he had a problem getting Hayden back into the Public Health System so he could have his operation.

Finally, having found someone who could see the subtle changes in Hayden, they had the confidence to ask a doctor to look at the MRI Sarah had done in 2005. They were told back then that there was nothing wrong with Sarah's neck, but Jenny's experience told her that doctor was wrong. To their relief, the new doctor identified instability in her neck, and thankfully they had time on their side to deal with Hayden first.

In 2007, Hayden was admitted to Auckland hospital for fusion and decompression of his cervical spine. However, that came with another set of problems. There were real issues acquiring operating time because Hayden had seen the doctor privately, and so getting back into the Public Health System created a nightmare. The more time they spent waiting for a date the more urgent the case became. Hayden was having balance issues. He was losing the use of his arms and spending more time in bed as life was becoming too difficult for him. Finally, they convinced the doctor that the situation was really urgent. They ended up having just a twenty-four hour notice to get to Auckland and present at the Emergency Department where there was an orthopedic team waiting for them so that Hayden could be admitted to hospital to have an operation the next day. They felt like they had to go in through the back door to get the lifesaving surgery!

He was fused from C4-T1 at the front of his neck and C3-T3 at the back. He wore a neck and upper body brace for six months to give him added stability and allow for the healing process to take place. Jenny thinks also that Hayden was really anxious about the instability he had been experiencing all those years and was not confident to move and sit up by himself. The operation took quite a lot out of Hayden, and it took longer for him to recover. However recover he did, and for a little while, he had really good health.

By April 2008 Sarah was having some real problems with severe headaches that would not respond to medication. So she was rushed to Auckland hospital for a Cervical Fusion. What they had not realized before then was that her skull had slipped down the odontoid peg, which meant that the peg was hitting the brain stem every time she moved her head. That explained why she was having such severe headaches.

Sarah was a real trooper. She was so determined to get up and get moving and put everyone to shame in the high dependency ward. She did an amazing job through her recovery process and was very relieved that the headaches became a thing of the past.

What bought a smile to their faces was that the surgeon had asked the original doctor (who had read Sarah's MRI and had told them there was nothing wrong) to assist with the operation. For the second time in their journey with Mucolipidosis, they received an apology. This doctor was totally shocked at what he saw during the operation. He said he had totally underestimated the extent of the damage ML could do. He thanked them for the experience of being able to help operate on such a complex condition. They were delighted that they were able to teach someone about Lysosomal Diseases. After these two operations and many visits to clinics in Auckland, Paul and Jenny felt like they were living at the hospital. They know the road from Tauranga to Auckland like the back of their hands.

In 2009, Hayden was again rushed into hospital for a full spinal fusion. By the time they found a date that suited everyone, Hayden was beginning to lose control of his bowels and bladder and again he experienced balance issues. However, once they were able to get this operation sorted out, the surgeon was a pro at obtaining operating time from his colleagues.

Hayden had a full spinal fusion both front and back from T4 to Sacrum. This surgery gave Hayden back his quality of life, with headaches becoming a thing of the past. He has learned how to move and manage having a stiff spine, and they saw utter determination from Hayden that he was going to fight through the recovery period and do what his sister did - return home and get back to normal quickly.

They found it amazing seeing him back in his chair. He looked as though he had grown in height overnight; of course he had, with his spine being straightened and now in a normal position, he was no longer hunched over. He was able to regain a good appetite and eat a reasonable meal, whereas prior to surgery eating was difficult with all his organs being pushed up under his ribcage.

The Noble family will forever be grateful for the care, dedication and compassion of the surgeon who listened to their concerns and successfully operated on Hayden and Sarah.

Huddy and Sammy Anthony

Sammy had several Urgent Care visits, all for respiratory distress from asthma and/or related to colds. They didn't go to the Emergency Room, but to the clinic where they administered nebulizer treatments (prior to them acquiring their own unit, which they felt was a blessing!), and they prescribed steroids each time. Liz well remembers the feelings of panic that accompanied those asthma attacks for both her and Sammy.

Callie Nagle

In 2005, Callie had a bowel obstruction and needed emergency surgery. Debbie took her to their local hospital where they did an ultrasound and decided Callie needed to be taken to a larger hospital by ambulance. This was by far the scariest experience they had up to that time. It was around two o'clock on a Sunday morning when they arrived in that Emergency Department, but they weren't seen until about four o'clock. Callie had been vomiting off and on for a few days, even missing the last day of school and the social dance. Her pediatrician had thought that Callie had a virus, so no one really knew what they were dealing with, and they were all exhausted. After Callie had a CT scan, it was determined that she had a bowel obstruction, and needed to have a naso-gastric tube inserted. Debbie was horrified! An NG tube goes into the nose and then down into the stomach, in order to help drain the toxins caused by the bowel obstruction. Two nurses attempted the procedure on Callie while Debbie sat in a chair silently praying. Suddenly Callie was trying sit up and started screaming and all Debbie could see was her face, neck, hair, and pillow covered in blood. One of the nurses asked if she had any bleeding problems, which she didn't, to her mother's knowledge. It was obvious that both nurses were rattled. Frightened and on the verge of hysteria, Debbie jumped to her feet and told them to stop. Poor Callie was absolutely terrified while Debbie was alarmed by the amount of blood she saw! She went out into the hall and asked to talk to the doctor who ordered this. When she appeared Debbie demanded, "What the hell just happened in there?"

The doctor said that probably the tubing they used was too big. Debbie said they should not attempt that again, even with a smaller tube. No more! But she wondered what was going to happen, because she knew the NG tube was crucial in removing the toxic fluid she had been vomiting for days.

Debbie was waiting for Rich, who was on his way to the hospital, when two surgeons approached and introduced themselves to them. They explain that Callie needed immediate surgery and one doctor very calmly told her that he would put this tube in her nose, but would use a spray to numb her. He promised to stop if she asked him to. They all went up to a room for the procedure, when the surgeon arrived and said that they were going down to surgery immediately instead, and he would put in the tube while she was sleeping. Debbie thanked God; her prayers were answered. Rich arrived in time to see Callie before she went in for surgery. It was late Sunday morning by that time.

About half an hour after Callie went into the OR, two anesthesiologists came out and said, "We had fireworks in the operating room." They explained that they had trouble accessing her airway. That was the first time Debbie and Rich had ever heard about that problem. Callie had six previous surgeries without any airway issues. Based on that situation, Callie was kept in a drug-induced state. She was in Intensive Care for the first few days after surgery. They didn't want to remove the breathing tube for fear that the airway would collapse once the tube was removed. So she was intubated and still had the NG tube in her nose for several days. They kept her heavily sedated and tied her hands so she wouldn't try and pull the tube out, which many patients naturally try to do. It was unsettling to see the emergency tracheotomy kit over her head. They said they would have to use this if that tube came out and her airway was obstructed. Callie's stressed parents were very relieved when they were able to remove the breathing tube without a hitch.

The surgeon had found that Callie had an internal hernia and adhesions, and said that he had never seen an internal hernia like that before. He claimed he named it the 'Callie Hernia.' She was very lucky that none of her intestines had to be cut or removed. She remained in the hospital for about a week and recovered nicely. Her parents were also relieved that the pain team was on board, and that made a significant difference during Callie's very painful abdominal surgery.

Debbie stayed many nights with her, and agonized over the decision to finally go home to get a good night's sleep, but Callie said she was fine with it. She really was a trooper, considering that previously she would never have wanted to stay alone in the hospital, and she recovered very nicely once she was able to return home. Debbie credited the power of prayer!

Jennifer Klein

Jenny's close call with a life- and-death situation from a serious pneumonia, while fighting an infection, is covered in a previous chapter. Since then she has had numerous trips to the Emergency Room, some being due to respiratory infections requiring immediate treatment.

She has also suffered very painful intestinal blockages. The first time she visited the ER for that problem was when she was eleven years old. Another reason for her ER visits has been a hip that has broken a number of times, due to the way the disease has ravaged her bones.

Andre Andrews

Jane has found that educating the people in the medical profession, as to what Andre's disease is all about, to be one of the hardest parts of dealing with his medical problems. As an example, when Andre had a minor asthma attack resulting from exposure to second hand smoke, Jane took him to the emergency room. The staff there went ballistic, even though Andre's records were there. They were there for about twelve hours overnight. The Cardiology staff was called in since that particular group had not seen anything like they saw on his tests, and they had major concerns. They wanted Jane to take him back to the clinic, but they finally spoke to his cardiologists who updated them on his condition and things went back to as normal as they could be for Andre. They indicated that there should be a note on Andre's file that alerts them of this condition so they won't panic.

They've had a number of visits to the ER through the years, mostly due to respiratory distress. Despite the fact that the notes are in his file, they always seem to panic there due to the fact he usually has high respirations and heart rate, and they simply don't understand the nature of his rare disease.

Spencer Gates

Andrea said that Spencer's first bout with pneumonia was not pretty. In the beginning, the eight-year-old boy simply said that he did not feel good. Later he spiked a high fever which started a cycle where it would break and then return. After three days, Andrea was ready to take him to the doctor when it broke again. The doctor said as long as the fever was gone she didn't need worry. So, that weekend they went camping, only to discover that they made a bad mistake! Spencer had diarrhea over the weekend and by Sunday he was wiped out. Andrea took him to the doctor Monday morning, and the doctor took one listen to his chest and said he had pneumonia. She sent them to the hospital for an X-ray and said she would call later. The X-ray technician told them not to leave because she was calling the doctor. Andrea felt a little nervous. A few minutes later the doctor said that he did indeed have pneumonia, so he would need an antibiotic. Relieved to learn that Spencer wasn't going to be a guest of the hospital, they went home.

Two days later he was not any better. They went to see the pulmonary doctor. Their regular doctor was not there, so they saw a different doctor. He listened to Spencer's chest and said he would be right back. A second doctor appeared, and also listened. Then she said that they would like to keep him in the hospital. Up to then, Spencer had been very blessed with being healthy. When he heard, "stay in the hospital," he panicked. Andrea asked if hospitalization was necessary, or if they could try something else. They agreed to let them try treatment in the ER. A little while later, they started an IV and oxygen. Seven hours after that they felt they could go home. They went on to see one doctor or another almost every day for three weeks. Over that time they almost had him as their guest at the hospital, but Spencer was relieved that he wouldn't have to stay there, and finally the pneumonia was gone. The many prayers that had been said for Spencer were appreciated by his parents.

Just before Spencer turned nine years old he woke up in the wee hours of the morning with a stomachache. Andrea could tell that his pain was such that he needed medical attention. He didn't object when she said she was going to take him to the hospital, so she knew he was really hurting.

Only a few days previous to that, she had learned through the ISMRD group that there was a medical paper with advice for anesthesia for the difficult airways of ML patients. She printed two copies and took them with her. Once at the hospital, she became worried when the staff told her that she didn't need to worry; everything would be fine. When Spencer's parents refused to sign consent until they personally met with the anesthesiologist, the staff took notice. Thankfully, it was the same doctor who administered anesthesia when Spencer received his port. He said that they had made him read something that time. They explained that this was a different letter and that they would not consent to surgery until he read it.

An hour and a half after Spencer went into surgery for an emergency appendectomy, the anesthesiologist told them that Spencer was fine and that they were right. Spencer has a very difficult airway. He ended up using the fiberoptic intubation after trying a breathing tube with no luck. He said that it was good that they insisted on having the anesthesiologist read the paper, and be aware of Spencer's condition. Andrea smiled and said, "Yes I know."

Spencer did well in the hospital after a few days. They were all happy that he was released from the hospital in time to be home for his birthday.

Sergio Cardenas

Sergio has had a few ambulance rides, but Maria Elena doesn't think they are worth mentioning. And, she is very thankful to God that he has not had to visit an Intensive Care Unit.

Zachie Haggett

On November 28, 2010 there was a loud thump at four o'clock in the morning that Brenda knew could only be one thing - Zachie had rolled out of bed and hurt something. Within forty-five minutes the squealing and crying started. She gave him an analgesic and tried to get him back to sleep. Given the history with Zach's other hip, all the signs were there. When a very pain-tolerant child is screaming and crying in obviously excruciating pain, his parents realize the situation could be a serious. A stronger medication helped him to become more comfortable, but didn't alleviate their fear that he had another broken hip. Although Zachie was more interested in being with family on Thanksgiving, his parents spent the morning mulling over all the options open to them and deciding what course to take. They thought they could keep him comfortable until they could see the specialists on Monday, but as the day went on, it became clear that wasn't an option. They conceded they would take him to the ER in the morning for X-rays. He demanded to go to his aunt's for dinner. When her brave little boy said, "First I was sick (a strep throat for three weeks), and now I broke myself; life is rough!" Brenda cried.

Realizing that Zachie was still in severe pain, they didn't wait until the next morning, but headed into the ER after dinner. They took X-rays and then a CT scan and saw no obvious break so they discharged him although he was still screaming still when moving his hip.

Being dissatisfied with that diagnosis, they looked further for help. It was not until December 10th that the orthopedic doctor, after scouring the scans and X-rays, confirmed there was indeed a small fracture on the backside of his hip. Zachie was to spend many weeks keeping weight off of that leg.

After months of waiting for the hip to heal, it was decided that a screw would be needed to keep the fracture from further damage. During that operation everything went well, and Zachie had an uneventful night. They were headed for discharge the following morning when he began wheezing. Within a few hours his quiet whistle became a loud roaring wheeze. His nurse asked if that was normal breathing for him, and learned that it was not. In a matter of minutes his breathing became so loud that he could be heard from the hallway. It was quite clear he was rapidly becoming worse! She made an immediate call to the resident who, before walking in the room, called for the rapid response team of nurses, residents, and emergency techs on the floor. Within seconds there were about a dozen people standing around Zachie trying to figure out what to do. First they tried the type of breathing treatment he usually had at home, but after five minutes (while his parents were feeling very anxious) that wasn't having any impact. The next course was an epi vapor used for allergic reactions, and he finally started to respond. During this time was Zachie crying and saying, "My lungs hurt! I can't breathe."

Brenda and John were surprised that he even knew the word lungs. They were very sure that God had kept them at the hospital long enough to be in the right place when Zachie had the attack. Had they left, the outcome might have been much worse. The breathing treatments were continued every two hours until almost midnight, and after that, every four hours. He slept well and sounded much better by the next morning, so the on-call ortho and respiratory team released him. The family arrived home by 1pm, and in pure Zachie fashion, he was rocking out to the music on the stereo by 7pm.

CHAPTER ELEVEN

FACING A MOST UNCERTAIN FUTURE

In recent years, groups have joined together to bring about awareness on Rare Disease Day. That happens only on one day in the year, despite the fact that these families are living with this every single day of the year. It is not something that can be set aside, and picked up again, now and then. The challenge and uncertainty is always there. They are always living on the edge; always wondering what they will have to face next. Most people make plans for the future. Yes, they are aware that many extraneous circumstances can arise to cause them to change their plans, but they also feel pretty confident that they are able to make those plans without expecting such interference. It is hard for them to imagine what it is like to live day after day with a progressive and fatal disease; not knowing what short range or long range plans can be made with any certainty.

Parents want their children to be realistic about their prospects. They deal with heartbreak as they see others move on with their lives, while many with rare diseases will always be dependent on someone else, no matter how hard they strive to be independent.

There is a certain amount of grief about the lost dream that families must face before they can move on to living a day at a time, as best they can, while dealing with all of the problems that they must confront. If one spends too much time grieving what is lost, every little joy of the moment will be missed.

Kelley Crompton

I prayed for acceptance and my prayers were answered. I taught Kelley those same prayers when she was young, and her faith became very strong as she grew. For many years her future didn't look much different than it did for others her age. But, as time went on, and the disease started to make its mark on her body, she found it necessary to make adjustments. Her life plan became a series of adjustments. She did forge ahead, and she was able to work for a while. She earned a college degree, and when the time came that she could no longer work, she decided to volunteer. Kelley made the best she could of the life she had, with her ultimate goal to be in Heaven with God when He called her home.

Bob and I were to discover that our retirement years became very different than those of many of our friends. Yet, in many ways our lives have become so much richer, because we have been able to witness first-hand the way others have accepted their challenges and worked toward making life better for others.

Kelley is no longer with us, and I do believe that she is with God. She met her challenges with such grace and acceptance that I want to honor her by remaining involved with the ISMRD and the MPS Society, and helping to raise funds for research. I believe that there is hope for those who still suffer by the work being done today in the medical community and at the Greenwood Genetic Center, in particular.

As I have observed other families dealing with children who have rare debilitating diseases, I have noticed that the parents not only worry about their children, and the future of their children, but they worry about their own health. They feel it is imperative for them to be healthy in order take care of their children. Yet, it seems that most of them have to deal with many health problems themselves. This is not only a worry for them, but for the children for whom they care. They put off admitting, even to themselves, that they are not feeling well, mostly because they don't want to worry others. Yet, we all have human bodies, and no matter how strong our will may be, or how fervent our prayers, we have times when we have to admit to being in pain, or being unable to do all that we think we should be doing. It produces a frustrating situation.

Autumn Tobey

It is with apprehension that the Tobey family looks toward the future, anticipating more surgery for Autumn, as the disease continues to assault her body. In addition to all of the bone and joint problems she has had, the doctors have found some issues with her heart.

Pam is determined to do everything she can to bring about awareness of ML III and conduct fundraising activities, while also working on the ISMRD Board. Pam has made a CD of spiritual music, with the proceeds going toward research. She will continue to sell jewelry, cookies, and plants, as she has done in the past. She hopes to make enough to have seed money to sell chances on $100 worth of gasoline or groceries, whichever the winner wants. The little town in which they live is too small for her to have a big event, and they have a lot of fundraisers locally for different individuals with health problems. Being aware that the requests can get too taxing on the people, she searches for creative ways to reach out for help.

Allison Dennis

Trish and Rich have found that having a child with a rare disease is hard, since there are no firm answers about what to expect. Each day they feel like they are walking on eggshells, not knowing what the next day will bring. They battle constantly to fight for Alli's needs and find the doctors and specialists to treat her. They try to live each day with zest to make plans for a future that may not be, but behind their smiles there is pain. Only those who really know them understand what they are feeling.

Alli's parents try to make sure that their daughter has the opportunity to experience all the wonders in life that they can give her. Years ago they made a choice that they would focus on the quality of life not the quantity. They feel that Alli has been fortunate and will continue to be so with the help of all the wonderful people they have met along her journey. They strive to make sure that her needs are met, so that she will experience as much of life as they can make happen. Trish believes it was the uncertainty that caused Rich to have a breakdown when they first learned of Alli's disease, and also the same stress that caused his stroke in 2010 and his heart attack and triple bypass in 2013.

Seeing their child suffer and being unable to make it better and take the pain away, causes them constant worry and pain. It all takes a toll on their health and wellbeing. It has also caused some isolation from friends and family, as they are unable to participate in many of the activities that others would like them to.

Their priority always has to be Alli and her health and wellbeing. But Trish sees a positive aspect to the situation, in that she believes it has made them all better people. Alli's brother Nathan works full time, but on weekends, and often through the week as well, he volunteers for the disabled accompanying them in the community. Nathan loves his volunteer work and Trish believes that it is because of his sister. She feels that, while they always have that constant shadow of fear and dread of what tomorrow may bring, they also have the love and the knowledge that, because of Alli and the dreaded ML III, they have all become more compassionate and understanding. They are grateful to have met some truly amazing people. Alli seems to have an ability to reach out and touch the hearts of strangers. Trish sees that in the other ML children that she knows, and believes it to be a blessing and a precious gift. While appreciating life and all the gifts it has to offer, the family never, for one moment, takes anything for granted.

Despite years of pain, with few ways of relieving it for any amount of time, Alli usually looks forward to the next activity she has planned. She likes to read, watch movies, bowl, and visit with friends. She would love to be well enough to go back to school to continue her education, and feels that she would be able to do much more if, as she says, "ML III didn't wreck my body."

Alli does whatever she can to appreciate all of the things she can do, instead of complaining about the limitations the disease has put on her life. She knows that she is not alone in her fight, while she prays that someone might find a treatment that will help her, so she won't need any more surgery. The cardiologists have told her that they could help her heart with surgery, but she has ruled that out. She feels that she has had enough surgery for anyone's lifetime. Her parents have been respectful of her decision.

While the pain she suffers causes frustration, in that it robs some of her energy, she refuses to feel defeated while requesting prayers on her FB page. The 'Brave Warrior Princess' has the determination and will power to live longer than people who don't have the kind of strength she has. She knows that she has met some very wonderful people she never would have known if she didn't have ML, while she also wishes there was some way to raise more awareness and to raise money for the scientists and researchers to find a cure.

Hayden and Sarah Noble

As the Nobles look toward the future, they see both opportunities and challenges. Sarah has been able to follow her artistic dreams of painting. During 2010, she entered the Genzyme Expression of Hope Stage II contest and won the Featured Art Award. She was so very proud, and she has gone on to sell several of her paintings. The family feels that it is wonderful to have some of her paintings hanging in their home. On the other hand, they expect that Sarah will need more surgery at some time. One shoulder has been successfully repaired, but the other one still causes her pain.

Sarah has joined Hayden in the Special Olympic Indoor Bowls Team. Hayden has been a member of the Tauranga Indoor Bowls Special Olympics team for many years and has his heart set on qualifying for next year's games. He is very proud of his gold and silver medals. He is an avid sports follower with his heart given to the NZ Warriors league team. At the age of eighteen

Hayden was made an honorary Vodafone Warrior - an honor not given to any other New Zealander.

However, because no one knows what to expect as the disease continues to progress in both Hayden and Sarah, the family must stay on top of any symptom that arises in either of them, with frequent visits to a variety of physicians, so problems can be addressed before more damage is caused.

Huddy and Sammy Anthony

Liz acknowledges that death seems like more of a reality to a family dealing with a terminal, progressive disease than it does to a family with healthy members. However, she sees that the truth/reality is that death could occur at any moment to anyone by a variety of causes.

The way she deals with this disease was and is, "One day at a time," and "My grace is sufficient for you," 2 Corinthians 12:9. She has many, many favorite underlined Scriptures, including:

Isaiah 43:2a, *"When you pass through the waters, I will be with you; And through the rivers, they will not overflow you."*

Psalm 139:13-16 is written in calligraphy in her Bible's flyleaf: *"For You create my inmost being; You knit me together in my mother's womb. I praise You because I am fearfully and wonderfully made; Your works are wonderful, I know that full well. My frame was not hidden from You when I was made in the secret place. When I was woven together in the depths of the earth, Your eyes saw my unformed body. All the days ordained for me were written in Your book before one of them came to be."*

Psalm 146:8 *"The Lord raises up those who are bowed down."*

Also: 1 Corinthians 15:43-58; Romans 8:11; Psalm 16:11; Psalm 18:30; I Peter 5:10; Philippians 3:20,21; I John 3:2; John 10:11-30.

Sammy's favorite verse was Psalm 3:4 – *"I was crying to the Lord with my voice, And He answered me from His holy mountain."*

Callie Nagle

Debbie believes that in dealing with the diagnosis of this cruel disease we all grieve, and go through a mourning process of what will never be, and we all have to find a way work through it. She tries not to think about the future too much, although she finds it hard not to do just that. She tries to focus on the fact that only God knows what is in our future and puts her faith in Him while praying for strength and guidance. Debbie acknowledges that it is not always easy to take one day at a time, but it really doesn't serve any purpose to do otherwise, other than to torture ourselves.

She is sometimes amazed at how much Callie has to endure, but she's so grateful for precious moments like when she hears her singing, or laughing, sounding carefree. If she allows herself to get stuck thinking about the future, she feels anxious, so she tries to focus on the milestones and the simple joy and pleasures that our children give us - which is a lot! Debbie is all for giving Callie the best life possible, and having those special celebrations on her birthday. That's when those special friends and family will come through.

Callie has had three cardiac MRI's over the last five years. The doctors thought she needed her aortic valve replaced, but then a third MRI showed a clearer picture, and she has been holding on; things have stabilized. She also has aortic regurgitation and mitral valve leakage, for which she is taking medications. Debbie prays that Callie won't need surgery for this issue, while she tries to stay off an emotional roller coaster.

Sometimes when she looks at her daughter, and all that she has to deal with, she compares her to other people thinking to herself that she doesn't know if the others would be able to deal with it. She believes that God has a plan for each of our lives - no one said that it would be easy waiting on God and His plan - that's where faith comes in. She tries to remember that God is there - always listening.

Jennifer Klein

Jenny is well aware of the way in which her disease is progressing in her body, but she stays positive when she thinks about her current state, as well as her future, knowing that there really is no way to accurately predict what will happen in anyone's future. She is attending college majoring in biology and psychology, with plans to continue on to medical school and eventually work with children with rare diseases, either in a doctor's office or a clinical lab. Jenny's life with ML III has taught her to be strong and to not take people, things, days or even her life for granted.

She really doesn't like to think that her mother is worrying about her. However, as much as she dislikes asking for help, she is willing to do so, when necessary. Each day Jenny finds a way to laugh and smile.

It troubles Terri to see the pain that her daughter suffers, and she will go anywhere and do anything she can in order to help alleviate that suffering. She left her lucrative work as Real Estate Broker when it became more important to her that she put all of her efforts into working toward trying to find treatments for Jenny and others like her, with the ultimate goal being the discovery of a cure.

Andre Andrews

Once Andre reached his eighteenth birthday, a new set of obstacles came into play with regard to treating him as an adult versus as a child in a pediatric environment. Jane felt like she was caught off guard with that issue, since she didn't see it coming.

However, everyone in the Genetics Department appears to be on the same page. They know Andre and understand what he is dealing with, so they tend to go out of their way to make sure that all of his needs are addressed. They have assured Jane that he will be provided for, as they will be able to continue there at least until he is twenty-five.

Both Jane and Andre are looking at his future in the same way they have everything else in his life – with a great deal of faith in God, they are taking it one step at a time.

Joey Nagy

Joey completed his high school classes 2013, and has gone into the Secondary Transitional Experience Program which is offered by the state of Illinois to those with disabilities until they reach the age of twenty-two. In this program, Joey will receive guidance and career counseling, opportunities for career exploration and job training, and placement from rehabilitation specialists, with the goal of finding a job and living independently.

Linda believes that whatever Joey's future holds, it will be one of tremendous love and support from a family who loves and accepts him for everything that he is and everything he does. He has made their lives so much richer. He came into their lives so that they would learn to 'stop and smell the roses.' He showed them what real love means; to love unconditionally, and to look beyond their own selfish needs and wants. For this gift, Linda will be forever grateful. She believes things happen for a reason and that God doesn't give us something that we cannot handle.

At this time, Joey just basically goes to his yearly specialist appointments. While the doctor recommends physical therapy, Linda is not a big fan of such. She believes the best therapy he can have is to keep moving on his own, so she pushes him, as she tries to keep him out of the wheelchair as much as possible, to make him walk.

As with many of the other families, they are looking toward the future by deciding to take it all a day at a time.

Anna James

Jackie acknowledges that it is necessary to be prepared and look ahead to make sure your child is secure and confident in their future, but there is a point where she has to 'disconnect' from facing the long term future for Anna.

She realizes that it may be an escape mechanism, but she simply doesn't dwell on the fact that Anna has a fatal disease. Although she can see the progression of the disease ravage her body and all the difficulties she already faces, she simply cannot look forward too much. She has to stay in the present. That doesn't mean that she doesn't look forward to make preparations for her care, such as what Anna can do for a job, and what she and Brent need to do to ensure their daughter receives the services she will need.

Their day to day life is very different from most families. Anna is unable to dress herself, bathe herself or use the bathroom without help. For Jackie, there is no jumping out of bed and rushing out to the car to work or wherever she needs to go. It usually takes a good half an hour to get Anna up, bathed, changed and then, of course, a little more time to fix breakfast. Bedtime takes a little longer too, as Anna needs to be changed and put to bed.

They consider themselves lucky that Anna is a very happy young lady who loves to read, play games, watch movies and generally keep herself amused during the day. While living in an old Victorian house (bought before the days Anna was in a wheelchair) they struggle with getting Anna up and down the numerous stairs and in and out of the bathroom.

All in all Jackie believes that having a child with special needs has opened her eyes to see just how precious life is.

Spencer Gates

Spencer is a high school student who is working hard to attain a good GPA because he wants to be able to go to a good college after he graduates from high school. He is a good student, who helps his sister, Sydney, with things like math, while she helps him with some of his physical needs.

Despite the challenges and pain that Spencer deals with, he looks forward to the day that a cure will be found. Meanwhile, he maintains a very positive attitude, and says that he won't let this disease stop him from achieving his goals in life.

As they face the future, Andrea feels it is very important for them to stay connected with other ML families through ISMRD and the MPS Society. She knows that fundraising and awareness campaigns are essential to working toward a cure, and believes the members of these groups have made a positive difference in Spencer's life.

Sergio Cardenas

Maria Elena thinks about Sergio's future all the time. Since Sergio was born, she would go to the park so her other children could play, while Sergio would enjoy the fresh air and sun. She always tried to enjoy those moments, but seeing all those beautiful children playing around, and moms there with healthy babies, made the reality of what she was facing even harder. She has never complained about what God has given her, but has always asked God to make her strong enough handle whatever He sends to her. However, she is a woman who doesn't accept "no" for an answer... or at least not until she fights all of her battles first. Those visits to the park made her feel how weak we can be, and how small we can feel. She saw the other mothers helping their children just one or two times to accomplish a milestone like using the swing, going down on the slide, walking, or playing with a ball, and then those children just could to do that by themselves, and that always reminded her that it didn't matter how much time she took with Sergio, because he never could do anything like that.

School plays, concerts, and sports activities have always been very hard for her, because she saw those proud parents and she will never be one of them, at least not for her little one. She really is proud of Sergio, and all he has been able to do, but she would like his accomplishments to be the normal ones. And it breaks her heart to hear him say that he would like to be like the rest of the children, playing sports, playing an instrument and mostly being the same size as others his age.

Their plans for the future are to try to help Sergio to keep going to school until he can graduate. And if he is still alive, they want to invest money and open a restaurant to make him happy, since he is always saying that his biggest dream is be a chef and have his own restaurant with his mother by his side. Maria Elena has put aside her personal plans so she can take care of Sergio and give him all of her time and love.

Every year the family plans vacations with Sergio's needs and likes and dislikes in mind. The rest of their lives are planned exactly the same way. They include their other two children, but sometimes Maria Elena feels bad and guilty that the other children would like to do just the opposite for their vacation.

The family's plans for the future include sending their oldest son to college in 2015 and their daughter to college in 2016. They we would like to sell their home and move to a smaller place. All of their other plans revolve around Sergio and his needs.

It doesn't matter to them that some doctors have recommended that they look for a respite care center, or placement for Sergio in an institution; they would never agree to let someone else care for their little one. Maria Elena could not live with such an arrangement. She gave him life and she promises him that she will take care of him for all his life and be sure to give him the best quality life she possibly can.

Every day Sergio's mother thinks about his future, and if she tries not to, something will occur to remind her of exactly what his future will and will not be. And of course it hurts, and it is hard to handle. Mucolipidosis is a very bad disease and a big cross to carry, but she believes that we need to love our cross and carry it the best we can without looking for any pity.

As the family approached the summer of 2013, they felt like they were on a roller coaster ride with Sergio's condition, after having many visits with doctors. Some of them felt that he could benefit from open heart surgery, but after all of the doctors consulted, it was decided that such surgery would be a risk and something he might not even survive. They also checked his brain pressure, and would like to recommend replacing the shunt he has, but that too was ruled out.

All of this means that Sergio's heart function will continue to fail and he will also experience severe headaches and some seizures, due to the pressure in his brain. The doctors have suggested hospice care for Sergio, at his home, to help him to handle the pain and become as comfortable as possible while he is sleeping more and eating less. And everyone prays.

Zachie Haggett

There isn't a day that goes by that John and Brenda don't think about the inevitable. John refers to the disease as a blessing and a curse. They experience a fear that many may never know their whole lives. Yet, because of that, they try to live every day to the fullest and appreciate every single moment they have with Zachie, even the more complicated ones, of which there are many. Since they do not know how long it will be before this disease claims their baby, they try to go to bed each night grateful for whatever the day has given them and hope they will enjoy tomorrow.

In 2009, when sharing about another parent's loss, Brenda said that she knew the day will come that she and John will be in their shoes. Yet, Zachie will continue to be strong and she will cherish moments with him forever. It is no easy task when faced with the constant knowledge that the disease is progressing until the disease wins. It means we all face the thought of days being numbered and life truly being a gift, each and every morning!

Brenda has been chastised by some for being a realist and for being honest with Zachie about things that may be beyond his understanding, but she believes that it is not beyond him. She believes that he does understand. He knows what it's like to watch from the sidelines, he knows what it's like to say goodbye for the last time to friends who have become their family, and he knows that people are the most important part of his world. He has told his parents that they are all he really ever wanted for Christmas!

There are many questions that are inevitable parts of life with ML. What if this is the beginning of the end? What if he never is able to walk again? What if this? What if that? It is impossible to not fear what is next all the time. It has become a natural part of life for John and Brenda, but the fact that Zach suffers so much pain stings harder than anything else!

Brenda sees Zachie as her hero. Every single day with him is truly a blessing and every single time he smiles it is another sign that ML has not won yet! While he may make things challenging at times, and he may make things downright frustrating, he is a hero for continuing to fight to walk and learn and live every single day. Even when some people do not give him a chance, he still continues to keep going! There will always be people trying to put labels on others or put people in 'Groups' or make people the way they want them to be, but in the end, Zachie is exactly who he is supposed to be: the child God sent to Brenda and John to care for and to teach them how to be better human beings, stronger human beings, selfless and nonjudgmental human beings, and most of all loving; they have an unconditional love for who he is and how far he has come.

Brenda feels that her little angel continues to teach her more courage, more strength and more grace than she ever knew possible! One morning, after a doctor appointment, as Brenda was lifting Zachie out of the car and into his wheelchair, the boy set her heart straight when he said, "Mom, you know what? When life knocks you down, you have to get right back up!"

CHAPTER TWELVE

SAYING GOODBYE

To date, only two families participating in this book have lost children, but the others know that day will most likely come for them, unless a cure is quickly found. And because of the way this disease has made us a close family, we have all shared in the loss of many other families who have had to say goodbye to their children. There have been special blessings bestowed upon us by those who have gained their angel wings, and those of us who have been left behind have learned from them.

For this chapter, I have asked each mother to contribute her thoughts about loss to this disease in her own words.

Denise, mother of Kelley Crompton:

Kelley was born on the 4th of July in 1963. I loved the fact that my first born would share her birthday with a very happy national celebration of America's Independence Day. We always celebrated her birthday with Red, White and Blue, including parades, cook-outs and fireworks. Even before we were married, I told Bob that I wanted my first daughter to honor my maiden name of Kelley. Being of Irish heritage, I always had an affinity for St. Patrick's Day, and she and I made sure we wore green on that day, even if we weren't going anywhere.

It was, therefore, most appropriate that when her lungs could no longer function, God took her into His loving arms on St. Patrick's Day, on March 17, 2009. She had reached the age of forty-five.

We prayed for guidance from God throughout Kelley's life, and He always led us to the right people for the help she needed. We are indeed very grateful to many members of the medical community that she was in our lives for so many years.

In her quiet way, Kelley touched many lives, teaching all of us about acceptance, patience and fortitude. She showed us that it is possible to be pleasant even when in pain or gasping for breath. She taught all those around her that it makes no sense to feel sorry for yourself or complain about your situation. We are all here for a finite amount of time. Some of our children gained their angel wings at a very young age, and some have been trying to hold out for the record. They all have had a profound effect on many people. God has used them to help people learn some of life's valuable lessons. The age at which one dies is not as important as how that person lived life. Kelley believed that, if you can adopt a good attitude when you play the hand that has been dealt to you, if you ask God for help along the way, you will have the strength to face the challenges of life, whatever they might be.

Kelley's spirit will be with us forever. When Grandpa Kelley died some thirty-five years earlier, I didn't realize how much of his spirit would remain with us. The years have taught me that, although we no longer have their physical presence, the spiritual presence of those who have taught us much about living life in harmony with God's will, that spiritual presence will always be available to help us through any trials we may face. We never forget people like that.

While she was a "Kelley" in the tradition of her optimistic Grandpa, she was also a "Crompton." She had one physical trial after another just as her Grandma Crompton did; surprising us all time after time by pulling through crisis after crisis. But the time finally came when her spirit needed to be released from that body that so restricted her.

I have never known anyone who had as much acceptance of God's will as Kelley did. There are many things that she never understood, but she did accept the challenges she faced during life with grace.

Pam, mother of Autumn Tobey:

I had such great hope several years ago that a cure would be found in Autumn's lifetime... that if I worked hard enough and told enough people... that someone would know a celebrity that would speak for these kids and raise lots of money for research. That was a dream that never came true. As each of these sweet children dies one by one, I guess I would say that my hope is growing thin. And as this horrid disease strikes another part of Autumn's body, I have had to face a hard fact... that a cure will not come in time to save her. I grieve so hard for each child lost and each parent that has had to live on without their child. This just shouldn't happen. Our children are supposed to bury us. They should have wonderful happy pain free fulfilling lives like we as parents had planned for them - not for them to suffer all their lives and die too young without a chance.

I am a Christian, and I believe that God has a reason for everything that He allows to happen. I have been told by many people that know Autumn what an inspiration she has been to them in their times of hardship and struggles. Her smile through all the pain is incredible. I find myself begging God to come again and end this old world and take us all together before I have to give my Autumn up. I don't have many answers, but I do know that whatever happens; God is still in control.

To finish my thoughts and sum this all up - we are on a journey that we didn't ask for and sure wouldn't have chosen it, but if it had to happen to us I can say that we have, or I have, grown and met some of the most incredible children and their families that God ever put on this earth. I love them all like family, even the ones I haven't met in person. We all share a bond that can't be broken... BUT I HATE THIS CRUEL DEVASTATING DISEASE WITH ALL THAT IS IN ME!

Trish, mother of Allison Dennis:

I think that every parent's greatest fear is that their child/children will die before them. I remember when Alli was diagnosed we were told by her treating pediatrician that one day we will wake to find that she has passed in her sleep. Every day, for the past twenty-four years, we face that dread. We wake several times through the night to check that she is still breathing. Alli's health has deteriorated now for the past few years and she has undergone many spinal surgeries. We have almost lost her at several times. Alli has cardiac issues, as do all of our children with Mucolipidosis III. For the past twelve months we have been told by the cardiologist that it is now only a matter of time. I have to admit I am scared! I don't want to see my child die. But at the same time I know I don't want her to suffer any more than she needs to. How do we deal with this? Too hard! We have had long discussions with our son Nathan (Alli's younger brother) and he is fully aware that Alli's time will come much sooner than it should. Allison has told me she is scared and that she doesn't want to die yet. She wants to live her life without pain and enjoy it. If only it were that easy. Alli is now thirty and I know that we are more fortunate than others that she is still here with us. I don't think we will ever be ready to say goodbye. Every night we check on Alli several times as you do a small child. We know that one day we will find her in eternal sleep. That terrifies us. But for now we will continue to appreciate and make every moment count for as long as we can, enjoying the blessings and most precious gift that GOD has given us.

Jenny, mother of Hayden and Sarah Noble:

There are no words that will ever describe the loss of a child, and it's hard when you have two children that you know one day you will lose. I think it's even harder for the unaffected child who also has to watch his brother and sister become more and more affected by this terrible disorder. As parents, Paul and I have already gone through the possibility of losing a child. Hayden was eighteen years old, and he was so very sick. The doctor had prepared us for his loss. But how does one prepare for that? The thought of losing our children is abhorrent, but we know they are only leant to us for such a short time, and that we must make the best of every day we have them. We have been through so much loss already, firstly the diagnosis and then the loss of them losing their mobility, the loss of them not being able to achieve good grades in school, the loss of a father not being able to walk his precious daughter down the aisle to be married, the loss of not being able to be grandparents.

Mucolipidosis is probably the cruelest of all the Lysosomal Diseases, and the only Lysosomal Disease where developing a therapy looks almost impossible. My heart breaks every time I hear that another ML or Lysosomal patient has passed away. Each time it brings it back to us both that our turn will come someday. No parent should have to face the death of their children before them.

Liz, mother of Huddy and Sammy Anthony:

My husband always feared that our boys would end their lives because of pneumonia in the hospital hooked up to machines. That was not the case for our dear little Sammy. He ended his nineteen-year-old life at home unexpectedly but peacefully. It was his heart. When the paramedics finally arrived and we had a question of trying to revive him, my husband spoke these words that I will never forget, "Let him go." My mind knew it was right, but my heart could not agree. It was the most difficult thing in my life I had to do. "We don't know what the future holds, but we know Who holds the future." Yes, we believed the timing of our heavenly Father was right and perfect for little Sammy. After all, he had spent over five years mostly immobile, needing total assistance for all his daily needs. Still, he had the most cheerful way about him, and had a ready smile to give and a large sense of humor. He was a big sports fan, following with great loyalty WI teams in the good and the bad seasons. He did not have pain the way ML III children experience it.

This is a prayer I recently wrote:

Dear Lord,

Thank You that Sammy so steadfastly reflected Your quiet patience and long suffering. He mirrored Your humility. Sammy had great love for us, his family which he had, because You first loved Sammy. Sammy, as Your child, esteemed You, Lord, highly. He respected You and Your Word. He knew You were the Lamb of God. He simply accepted Your Word with Your commandments and promises. He took Your care for him from our hands and it was for him enough. He rejoiced in Your love for him! He grew in wisdom and knowledge of You and never tired of hearing Old Testament stories. Sammy liked hearing us sing hymns. He did not have much breath to sing with, but would usually sing the refrain of "To God be the Glory". We miss Sammy's presence.

We know Your Word is absolute truth as it says, "I am the way, and the truth, and the life; no one comes to the Father, but through Me." (Jo. 14:6) "And the testimony is this, that God has given us eternal life, and this life is in His Son. He who has the Son has the life; he who does not have the Son of God does not have the life. These things I have written to you who believe in the name of the Son of God, in order that you may know that you have eternal life." (I Jo. 5:11-13) "In My Father's house are many dwelling places;.. I go to prepare a place for you."

Debbie, Mother of Callie Nagle:

I live with the sad reality that Callie's life was robbed without her having had the chance to live a full life, a "normal" life, due to Mucolipidosis. But then her life was already robbed of living a full life the way Richard and I dreamed of, not to mention Callie's dreams. She still has dreams for her future and who are we to discourage or deny them? I try not to dwell on the sadness of her life and limitations, otherwise I wouldn't be functional. My faith has definitely sustained me and I will never give up sharing my faith with Callie, who sadly (at this time) thinks God is absent from her life. We know that's not the case. I've gone through many seasons of grieving through her life - along with her. She knows what she's been missing out on and feels so desperately that she's different, it just breaks my heart. My Callie is a spirited young woman and there's no denying what a spitfire she is! She's been an inspiration to so many; I only wish she believed this and could see herself as others see her. She is striving every day to live life to the fullest and amidst the sorrow we still have a lot of joy in our lives.

Many years ago as I often thought about her life and this horrible disease and what it really meant for her future, I came to the conclusion that her life could end in many ways as I saw many tragedies among young people whose lives ended "before their time." I found that I was dwelling on this and because this disease varies so much from child to child, no doctor could give us any information as to her lifespan, I realized I just have to treasure our time together. This was a realization that actually provided just enough peace and relief to stop dwelling on something I had no control over. God is in control. He has put so many people in Callie's path, she has touched lives and I believe this was His plan for her life.

Terri, mother of Jenny Klein:

The first time a geneticist told our family that Jenny had Mucolipidosis, he shared with me life expectancy, in years. I will never forget that day. It was a day that propelled me into a place of sorrow and it took years to see the light, get back on my feet emotionally and become the caregiver I was intended to be.

After diagnosis I went through the stages of disbelief, sorrow, and mourning. Eventually I came through stronger and wanting to make a difference for these rare diseases. I did not choose this journey for my daughter. And there was a day I realized, *I did not have a choice.* Jennifer does have Mucolipidosis, so what are you going to do about it? How are you going to help Jennifer have the best possible life?

My journey to understand the disease, the limitations my daughter would face in her life and to understand loss would constantly redefine itself with each passing year. It has been difficult to share loss with Jennifer and discuss friends that may have succumbed to Mucolipidosis or a related disease. If I had thought I could put days like these into a locked box, I was wrong. It has become critical emotionally to visit loss, remember someone's life and how they touched our family. I cope with loss by embracing the memories of that special child and to never, ever forget.

There are many days when we wish we did not have to see a doctor or visit a hospital with the many medical issues that come with having a rare progressive disease. We try to balance these days with something positive and plan trips around these visits. Instead of looking at each year with a medical issue progressing, we have come to understand that this is Jennifer's life and it is just a small part of it.

My thoughts and how I lead my life do not focus on loss. Instead, Jennifer has taught me to focus on life. Her strength, charisma, optimism and yes stubbornness have been successful at helping me remember our life here is not about the loss that will eventually come, but about how to find special moments in each day.

We were told Jennifer would live into her early teens, but we soon learned that this was not accurate. My hope is that while we are on this special journey, that I am given the opportunity to always be Jennifer's caregiver and that someone in her life she can count on to navigate these uncharted roads. I want to be able to reflect back and say, "We had no regrets."

Jane, mother of Andre Andrews:

Loss to this disease . . . . Once I was given what seemed like a life sentence, the world seemed to stop. I was told that my son's biological clock was ticking away and because the prognosis (based on the literature at that time) was only up to age 3. I was told to take a multitude of pictures and to give him all the love I could. He was getting ready to have a birthday that I had to try to not to believe that he may not see, and I went into overdrive to make sure that no matter what, he would have everything that life could offer. I was told by my mother-in-law at the time (who was a devout Christian) that his life was in the hands of the Lord and that "He" had the last say. I too, believe in God, and was convinced that God would have the last say and would just have to be ready. It was a little scary and each birthday was now a celebration of life. I was surrounded by friends with breast cancer and other tragic illnesses who were fighting just like he was and somehow had the strength to continue to run this race so we were not going to have a pity party but were going to live! I have received a lot of support through my relationship with the parents in the MPS society and the ISMRD and have made many friends. As a result, this was how I met you, and through our forums, conferences and friendships, I became aware that I was not in this alone and that there were others who shared this journey with me. I had to come to grips with dealing with the loss of this disease through the losses of many of the children who we have lost. There were some with whom I was extremely close and these losses were difficult because we spent some really valuable, as well as quality, time with these families and we still try to be there for each other knowing that one day, our time will also come.

I have learned to live each day as if there is no disease. It will always be there, lurking in the back of my mind and sometimes the thought overwhelms me and I cry. Some nights, I look at my child sleeping peacefully and hope that tomorrow will not be that day. I can't predict what that day is going to be like for me so I go on. I know loss for me will probably be devastating since this little boy, who has now grown into a young man, is such an important part of my life. The thought of him not being here is just as painful as it was losing both parents. Even though you know that day will come, you put it in the back of your mind until you have to deal with it. I know we have been on what is considered 'borrowed time' but I have been given a chance that most people don't have, and that is to let them see Andre for who he is and understand that he is really no different from you and me.

I want I-Cell parents to know that loss of life is inevitable but to feel encouraged at the idea that loss will come to each and every one of us at some point in time. I am thankful for every minute of his life and through all the sacrifices, I have come to grips with knowing that I have tried, and will continue to do the very best that I can to make his life as good as it can get. Tomorrow is not promised to any of us. I may even leave this world before he does, but for as long as I have him here, to bring me all the joy to my life that he has, I hope that when the day comes for me to deal with this loss that I can let it go and leave it in God's hands as He is the author and finisher of my faith.

Linda, mother of Joey Nagy:

I'll never forget the day we received the diagnosis. I don't think I ever truly knew the meaning of loss until that moment. It hit me like a ton of bricks. All my dreams and hopes I had envisioned for my son were suddenly gone. These are things one imagines even before the child comes into being. All parents dream of their children attaining a multitude of what life has to offer; among those are good health, happiness, family, education, love and success. You picture what your child will become as they grow and mature into adulthood. You hope for them to one day have a family of their own and give you grandchildren someday that you can enjoy. You are aware that not everyone chooses the same path, but you know the opportunities exist. For us, the reality of his diagnosis extinguished our aspirations for him.

Later, after everything had sunken in, you're faced with the reality of this super orphan disease. With so little population affected, little or no research is being done because of lack of funding. You are left with your only option; connect with others dealing with the same disease. We anxiously sought out such support. Through this journey, we have been fortunate enough to connect with many other families just like ours. We have gained a sense of belonging. Lasting friendships and bonds have been forged. We've experienced their personal losses and felt their pain. We've hoped together, grieved together and learned invaluable knowledge along the way.

Each time a family loses someone to this awful disease, we can't help but feel like we've lost a part of ourselves. It's especially hard when they are young children and taken so prematurely. I cannot fathom the loss that those parents must endure. I'm certain each death frightens all of us. As we mourn their losses, we contemplate our own child's future, but we convince ourselves that it will be years away. I think that is a coping mechanism.

I think for me though, my biggest anguish is to watch my son grow and not experience things like his older brother. His loneliness and isolation is heartbreaking. He tries to hide it and always puts on a smile. Despite his best effort, I know he feels sadness and disappointment. I know he longs for things to be different. He looks to me as his source of companionship; for which I am more than happy to provide. But, I can't help and wonder if this is enough.

I worry about his future and where it will lead him. What new tribulations await him? As a parent, again I find myself at a loss; how can I ease his suffering? Will I make the right choices for him? Will I be strong enough to accept the changes as they come? I constantly have to remind myself of this saying: "Worry does not empty tomorrow of its sorrow; it empties today of its strength." - Corrie Ten Boom

Jackie, mother of Anna James:

This is not a place I can really go to in my mind. While I know that it is something that will happen in the future, I think if I went to that place, it would completely bring me down.

As Anna gets older, I know in my heart that my time with her on this earth gets shorter, but I try to look at it as every minute I get with her is a complete joy to me and everyone she encounters.

I have had the honor and pleasure of meeting some amazing young men and women that have battled this disease and that are now gone from this life, and I'm not naive enough to think that I won't be dealing with the loss of my daughter one of these days, but my heart tells me to focus on what is now, not on what will eventually happen.

I'm not even really sure that Anna even knows that her life will be shorter than her friends and I'm ok with that right now. I want her to enjoy life to the fullest and not be afraid of what the future might or might not hold.

I won't lie or evade the truth from her, but she will know more as she is ready to hear it.

I realize that there are more surgeries, more pain, and more questions as we walk this path. But, one thing at a time. This journey is a journey taken step by step and both my husband and I will be beside Anna the whole way.

Andrea, *mother of Spencer Gates:*

What can I say about loss? Loss has many different meanings for me. It's something I think about every single day, some days more than others. It can be the loss of life or fear of that loss. For me loss can mean all the things Spencer may lose out on. I can still remember how excited we were to meet our first ML family. We had so many questions and couldn't wait to see what our son's future would look like. Spencer had only been diagnosed for a couple of years and he was doing well and showed little to no symptoms. My heart sank when we walked up and saw our first ML family in a wheelchair. I guess at that point I never even thought of Spencer's future involving a wheelchair. It was a loss I was not ready for.

Unfortunately that loss came when Spencer was in the third grade. Spencer had been dealing with pain for years and it was getting to a point that something needed to be done. Spencer's daily pain became so unbearable that he could no longer run or walk long distances. He ended up in a wheelchair to relieve some of his pain. He was heartbroken. The doctors told him that once he was older he would be able to have hip replacement and walk again.

I can still remember when Spencer went in for osteotomy surgery in 2008. He was recovering in ICU. Kevin's parents had come to see him and wanted us to take a break and get something to eat with them. We were leaving, but I wanted to just check on him one more time before we went to the cafeteria. But when I stepped back into the room there were two nurses by him and lots of machines going off and both of them talking very fast. When they saw me they just looked at me, then each other and kept on working. I'm not sure for how many seconds it was that he was, I don't know what to call it, but I will never forget that moment. I just remember the nurses looking up at me after they did their magic and the machines were quiet again and their faces said it all. The incident was over. I walked over and gave him a kiss and thanked God. That's as close as I want to come to loss with him.

Recently we were told that new hips were not in Spencer's future. When Spencer hung his head down and single tear fell, that was loss. But my heart truly broke when we got in the car and he said that he had been waiting all these years to walk again and now he will never walk again and no one will ever fall in love with him. Spencer had been waiting for the dream of new hips for seven years! To be honest we were dreaming right along with him.

So, what is loss to me? Loss is anything related to this terrible disease that breaks your heart or spirit. This disease has taken so many lives. It has also taken so many dreams, hopes, and spirits.

Maria Elena, mother of Sergio Cardenas:

When I was told about Sergio's diagnosis, my first question was, "What will be the next step?" And then I had many more questions, "How can I help him? How can we cure him? How do we stop this disease?" But after a couple of long minutes of silence, and then taking a big breath, the doctor explained in more detail about this disease. I can tell you that since that day, I knew that my first enemy in this battle was that Sergio could die soon. All those signs and symptoms were just how death is reminding us how close Sergio is to it every day. I try to talk with him about all my feelings always, to let him know how much I love him, how much I love taking care of him, how important he is to me and to all the family. I include my children in doing the same with Sergio, and talking with him about the fact that maybe ML will take him away from our side before we want, and that's the reason we need to live deeper every day, every single moment, taking lots of pictures, videos, singing, dancing, doing whatever we can together just to make him happy, and for us to feel as happy to share everything together while we are alive. And that includes also, farewells, like every night telling him how much I love him, and showing him all my feelings just in case he doesn't make it until the next morning. Maybe it sounds crazy or exaggerated but this is how it is to me. I do the same before he goes to school, and when he needs to go under general anesthesia. I tell him that if he never comes back I will miss him and love him forever, and that I will meet one day in heaven. And I also always tell him that if he is alone and feels something is happening and sees a light and/or an angel or something similar, then to ask Jesus to please wait and let him say goodbye to me.

I know that when Sergio dies, he will not suffer anymore, but, it will be so hard for us, especially for me, to say good bye to a lovely boy who is always teaching me to have faith, to wait for a miracle, and to not see all those beautiful dreams that Sergio never could achieve like going to college, finding his soul mate, getting married, having children, etc.

I know this moment will come, and I also know that it doesn't matter how hard I try, I will never be the winner in the battle against this monster. Sergio will no longer be suffering, but this disease robbed my son.

I think in this moment more than I like to, especially when he is under general anesthesia, or when I receive the bad news about another child who lost his/her battle against ML, and ask myself who will be next? Maybe Sergio? And the feeling is so deep, that I cry for hours. Knowing your child will die sooner because of ML never prepares me for that moment, and also I think I'm not prepared and I never will be prepared for it. But, I will accept it, the same way I have accepted every bad thing this disease has given to Sergio daily.

Every day I will use my efforts to find a way to let people know about ML and one day find a cure, a treatment, anything to stop this monster. The worst battle is that one that you never take on and fight in.

Brenda, mother of Zachie Haggett:

Nov 1, 2005: These past few weeks have been probably the hardest days since we learned of Zach's diagnosis and this morning we have yet another baby lost to this horrible disease. There is no easy way to explain how difficult it is to not only our family, but every family who deals with this death sentence every day. There are days and weeks and months where we get to live in a little bubble of normalcy and pretend that we don't even think about it and then comes reality crashing in and we are reminded of how extremely vulnerable our children are and cling to them even harder and cry even longer than the time before. In times like these, the family we have formed of affected children means more than words can express and yet at the same time steals another moment of joy and happiness, and a life that has been ever so cherished from the very first breath to the very last second. We hold our own babies even tighter and cry even harder than the last time.

This is, quite frankly, the hardest life I could ever have imagined and sometimes wish I never knew. Each time a child passes away, I feel another part of my heart ripped out and hope snatched away. I feel so guilty sharing these thoughts here and know many who will visit are here to get an update on Zachie but in a way, this is an update on Zach and all of the families we are so very close to because of these diseases; it is once again a jarring wake up call to many who only hear of the details from us from time to time. This is yet another reminder that we are still too far from a cure and desperately need help funding the researchers. No one should bury their child. No one should watch their child grow weak or filled with more pain. I am grateful sweet

Robin is pain free now, she suffered too much in her young life, and I am certain she is now having so much fun - free of this life-robbing disease and causing quite a hoopla up in heaven. It certainly warmed my soul knowing her mother was there to greet her and it helps me to know she will absolutely be looking down on us all, especially Zachie, and will one day be there to hold his hand as well. It is so hard to grieve these precious children without inevitably thinking about our own time. When will it come, how much longer do we really have with our own baby? It's a question and answer that none of us really know but it is definitely a question that some of us must ask too soon. It's just not fair. I know it's life but it is so damn painful and so completely heartbreaking. As I sit here contemplating my broken heart, Zachie is still playing next to me and for another day we are blessed to still be able to hug him and treasure him.

AUTHOR PROFILE

Denise Crompton and her husband, Bob, are the parents of four children, the oldest of whom had the ultra-rare disease of Mucolipidosis III (ML III). In 2003 Denise authored *Kelley's Journey: Facing a Rare Disease with Courage* to share the Crompton family's experience with others who were facing the same challenges they had been dealing with for many years.

Ms. Crompton's background, while varied, has always included working with people. She worked in the nursing field before obtaining a degree in social work, which led her to working with special needs adolescents as counselor and job coach; parent aide; substance abuse counselor and owner/operator of a specialized bookshop in Nashua NH, working with local counselors to carry books they recommended. In addition she held a Real Estate Broker's license in Massachusetts and New Hampshire, where she worked as a Buyer's Agent specialist.

Diagnosis: Rare Disease is the natural outgrowth of her expertise in caring for her daughter, searching for and finding help, never missing a day when she was hospitalized, and being "on duty" 24/7 for the last 5 years of her life, as her condition deteriorated until her death at the age of 45.

She has attended many conferences of rare disease groups, written numerous articles for their periodicals, and given a presentation to parents and doctors to assist in the development of positive relationships. For many years, she has corresponded with numerous other families, sharing their many trials and triumphs. Her desire is to help them, and all who suffer from rare diseases, by taking readers inside the very complicated world that a rare disease creates. She intends to donate most of her proceeds to medical research for rare diseases.

On Facebook?

Join the "Diagnosis: Rare Disease" Facebook community page and participate with other family members and friends. It's a great forum to share your experiences, post photos and much more.

https://www.facebook.com/diagnosis.raredisease

PRESS

Books that Change Lives

"I Am Not a Syndrome – My Name is Simon" by Trisomy mom Sheryl Crosier
Read the gripping story of baby Simon Crosier and his parent's fight for respect from the medical community. Born with Trisomy 18, a genetic disorder, Simon touched the lives of so many people. This is his story.

Proceeds from this wonderful, pro-life book supports *SOFT – Support Organization for Trisomy 13, 18 and Related Disorders.*

"Angel Gabriel - A True Story" by Joy LaPlante
Available in all e-book formats and paperback edition.
This is the true story in the aftermath of Comair flight 3272 which crashed in January, 1997. Don't miss this incredible story as told by Monroe, Michigan resident Joy LaPlante.

HEALTH / WELL-BEING

"Sleep Great for Life" by Rich Nilsen.
"This book offers the solution to one of life's greatest health risks."
Learn the 15 steps for a solid and secure sleep foundation, and then start applying the secret key to great sleep for life. "Sleep Great for Life" will help you overcome your insomnia so you can start reaping the benefits that come from a great night's rest. It available exclusively in Kindle format through Amazon for under $5, and in paperback direct from All Star Press.
This is a must-read for anyone who suffers from insomnia one or more times per week. This book is rated 5 stars on Amazon. Order today in Paperback or download the book to your e-Reader today.

"The Road to Recovery: Overcoming and Moving Beyond Your Grief"
This easy-to-read ebook is a tremendous comfort and resource for anyone who has suffered the loss of a loved one. It was originally written and distributed free for the Sept. 11th families. This book is rated 5 stars on Amazon. Download "The Road to Recovery" to your Kindle eReader today!

All Star Press publications are available in for Kindle readers at Amazon.com.

This is just a preview of the All Star Press book catalog. Order any of these publications at **www.allstarpress.com.** Thank you for looking!